Machine Age
To
Jet Age
Volume III

Radiomania's® Guide to Tabletop Radios
1930-1962
(with market values)

Mark V. Stein

Radiomania® Publishing
2109 Carterdale Road
Baltimore, Maryland 21209

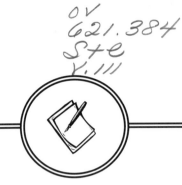

NOTICE

The market values indicated in this reference are based on a number of different sources. Actual prices will vary dependent on many variables. Neither the author nor the publisher assumes responsibility for losses which might result from the use of this book.

Published by
Radiomania® Publishing
Copyright 1999

ALSO BY MARK V. STEIN

Machine Age to Jet Age: Radiomania's Guide to Tabletop Radios 1933-1959 Vol. I
Machine Age to Jet Age: Radiomania's Guide to Tabletop Radios 1930-1959 Vol. II

Additional copies of this book may be purchased or ordered
from your local bookseller, or directly from the publisher
by sending $29.95 ($24.95 for Volume I, $27.95 for Volume II)
by check or money order payable to:

Radiomania® Publishing
2109 Carterdale Road
Baltimore, MD 21209 USA

(US priority and foreign surface shipping free)

E-mail: radioman@crosslink.net   Fax: (410) 466-0815
Website: www.radiomania.com

Future editions of this reference are currently in progress.
If you care to contribute photographs, advertising brochures or other items,
please write the author directly in care of the above address.
All contributed items will be returned upon request and contributors given
due credit in the acknowledgments

Cover designed by Jane E. Rubini
Book designed by Mark V. Stein and Jane E. Rubini

Dedication

For my girls,
Chloe and Isabel.
For tolerating all of those
dirty old radios in our home,
and for not pulling off
the knobs.

For your love and patience
I thank you.

ACKNOWLEDGEMENTS:

THIS BOOK WAS REALIZED THROUGH
THE COOPERATIVE EFFORTS OF MANY
COLLECTORS. THE MAJOR CONTRIBUTORS
ARE AS FOLLOWS:

NEW ERA ANTIQUES/STEVE CAIATI
DEALER AND COLLECTOR SPECIALIZING IN HIGH PROFILE ART DECO
RADIOS, TELEPHONES, TELEVISIONS, LIGHTING, CHASE, AND MORE.
WWW.NEWERAANTIQUES.COM

BRUCE EDDY
COLLECTOR SPECIALIZING IN UNUSUAL 1930S RADIOS INCLUDING
GLOBE DIALS, HIGH TUBE COUNT, MOTORIZED TUNING AND IN
1950S VARIABLE SELECTIVITY RADIOS
2521 W. NEEDMORE HWY., CHARLOTTE, MI 48813

RADIO RELATED WEBSITE CONTRIBUTORS:

THE FOLLOWING RADIO-RELATED WEBSITES HAVE LENT THEIR SUPPORT TO MY EFFORTS BY
GRANTING PERMISSION TO USE SOME OF THEIR IMAGES. ALTHOUGH I DID NOT USE AS MANY
AS I HAD ANTICIPATED, THE LEVEL OF INTEREST IN HELPING WITH THIS PROJECT WAS OVERWHELMING.
I ENCOURAGE YOU TO VISIT THESE AND OTHER COLLECTORS' WEBSITES:

HTTP/WWW.CROSLEYRADIOS/COM
CROSLEY COLLECTORS' SITE/JIM WATSON, SPONSOR

HTTP://MINDSPRING.COM/~CACUTTS/RADIO/RADIO.HTML
RADIO COLLECTION OF ALLEN CUTTS

HTTP://WWW.RADIOPHILE.COM
THE RADIOPHILE/JOHN PELHAM, SPONSOR

HTTP://WWW.GEOCITIES.COM/~JROSE3/
JERRY'S ANTIQUE RADIO/JERRRY ROSE, SPONSOR

HTTP://ANESTHESIA.MC.DUKE.EDU/RADIODAZE/
RADIO COLLECTION OF LARRY DOWELL

HTTP://WWW.GEOCITIES.COM/THETROPICS/2468/RADIOINDEX.HTML
CAPE OLD RADIO/RADIO COLLECTION OF CARLOS LAZARINI FONCESCA

HTTP://CPU.NET/CLASSICRADIO/
CLASSIC RADIO GALLERY/SCHEMATICS SERVICE/MERRILL MABBS, SPONSOR

HTTP://WWW.TRI.NET/RADIO/
ROD'S CLASSIC & ANTIQUE RADIOS/ROD ROGERS, SPONSOR

HTTP://ANTIQUERADIO.ORG/INDEX.HTML
PHIL'S OLD RADIOS/COLLECTION OF PHIL NELSON

HTTP://HOME.ICI.NET/~SFULLMER/SFULLMER.HTML
STEVE'S ANTIQUE RADIO PAGE/COLLECTION OF STEVE FULLMER

IN ADDITION TO THE ABOVE, THE FOLLOWING INDIVIDUALS HAVE ALSO
CONTRIBUTED TO THE COMPLETION OF THIS BOOK:

JENNIFER PURKIS, JON STEINHAUSER, STEVE CHAPMAN, DON HOWLAND,
AL BERNARD, LARRY DOWELL, RADIO BILL FROM DETROIT.

THANKS TO ALL!

PREFACE

Welcome to the third volume of Radiomania's vintage radio reference guide. Thanks to your continued support of this series. I am please to present to you a third and supplemental Volume Three, more than two years in the making. In compiling this third volume reference, I have tried to stay true to the intent of the first two in focusing on a single, high profile area in vintage radio collecting, the tabletop radios of the high style era beginning in the early 1930s and continuing through the 1950s. I have slightly broadened the time frame in Volume Three, once again, to include the years 1959-1962, so as to encompass late model tabletop radios which have recently begun to attract the attention of collectors. This reference serves as a supplement to Volumes One and Two, and is comprised of all new listings so that the radios pictured in the three volumes combined total over 7,000. This is, by far, the most comprehensive reference book on the subject and I am far from finished. That's right, Volume Four, in addition to other radio-related references, is in the works.

In compiling this text, I continued to use collectors and collector meets as resources for subject photography. In addition, through the generosity of advanced collectors such as Bruce Eddy and Steve Caiati, I have been able to tap large resources of original advertising and collateral industry materials to further broaden the scope. My hearty thanks go out to these and all hobbyists involved in the development of this book. I encourage others to share their resources with fellow collectors through the continuation of this series. Your thoughts, comments and contributions are welcomed.

Thank you and enjoy!

Mark V. Stein

PRICING

In compiling this book the question as to whether to include prices at all was one of major concern to both the author and contributors. As most collectors are aware, even the best price guide, if not initially flawed, is soon obsolete. The range of prices paid for most items is wide. One can pay anywhere from a few dollars to a few hundred for a given item, dependent on where it was purchased and from whom. Thus the task of establishing a 'market value' is difficult if not impossible. This problem given, it was still generally felt that the value of including prices, at least as a general market gauge, would outweigh the disadvantages of omitting them completely.

In establishing values a number of sources were utilized including auction results, classified ads, meet pricing, collector valuations and the author's personal experience having been both a collector and dealer for over fifteen years. Prices in this book represent items in average condition. Average meaning that the radio is intact but 'as found'. No repairs would have been made, either electrically or otherwise. If wood, the finish would be original. It might show some wear, but would be presentable in a collection once cleaned up. If plastic, the cabinet would be free from cracks and chips. Plaskon radios in average condition might evidence some minor stress lines but nothing which would detract from their aesthetics. Values followed by a '+' symbol represent rarer items, only a few of which have traded hands. In such cases the value listed represents an estimate of what a collector might expect to pay for an average example. The '+' symbol is also used to indicate baseline values for catalin radios, the price of which will vary widely based on color, condition, marbeling, and variations. This reference does not attempt to address the intricacies of valuing catalin radios, but merely to provide the reader with a baseline gauge.

A word about dealers and dealer prices: expect to pay a premium when purchasing from a dealer. The dealer offers one the luxury of eliminating the time consuming hunt through yard and estate sales, flea markets, antique shows and the like. It is he who goes through the trouble of rooting out those hard to find items. Ones which you might not happen upon except after years of hunting yourself. Dealers inventories represent long hours and related expenses and must reflect those additional costs.

FINE TUNING VALUES

To assist the collector in fine tuning the value of a given radio, some general rules of thumb are offered. Please remember that, as with all rules, there are exceptions to these. If you are uncertain about a specific item you will always do best to ask another collector whom you trust.

GENERAL CONSIDERATIONS

CHASSIS CORROSION: In buying any radio it is important to visually inspect the chassis, particularly if you plan to restore the set to working order. Minor surface corrosion is typical and acceptable to all but the most finicky of collectors and should not deter from the value. Extreme corrosion can be an indication that the radio was submerged at one time or at least has seen a lot of humidity. If so, the bulk of the internal components may need to be replaced. Even if you plan only to display the radio, if it is so damaged its value will be decreased by the fact that it would be less desirable to most other collectors.

RODENT DAMAGE: Most of us have, at some time, come across a radio with leaves stuffed into the nooks and crannies of its chassis and acorns which have mysteriously gotten underneath, or a set which has had every wire, both exposed and internal, chewed through. Such phenomena are caused by common vermin such as squirrels, rats and mice. At one time the radio was probably stored outside in a barn or garage and it happened to become someone's home. The damage caused by such occupancy can be severe. If you come across such a set and are tempted to buy it, inspect it carefully. The easily observed superficial problems may be but a sign of more extreme damage internally. Remove the chassis if you plan to restore it to working order. You may chose not to after further inspection.

TRANSFORMERS: A common problem, and one that may be costly to remedy, is that of the 'smoked' transformer. Caused by a short circuit or overload, damage to the transformer almost always necessitates its replacement. Fortunately, such problems are often easy to spot. Look for smoke damage to the chassis and on the inside of the cabinet. Also look for gooey stuff (which may have hardened over time) oozing from the transformer itself. Most collectors stay far away from sets with blown transformers if at all possible. Short of finding a junker radio of the same variety but with a good transformer, you'll need to buy a replacement with the same specs.   This can cost anywhere from twenty to ninety dollars assuming you are lucky enough to even find a match.

TUBES: Missing tubes are generally easy to replace, particularly amongst radios from this era. The resources are numerous. Your local club will probably be your least expensive resource (usually one to three dollars per tube). After that there are local repair shops, mail order companies and specialty houses. It is a good idea to know your resources when considering the purchase of a radio sans tubes. Also be aware that there are components known as ballast tubes which look like filament tubes superficially but which are really resistors. Count on about one out of every two ballast tubes needing replacement. Unlike filament tubes, few ballasts have survived. Replacing a bad ballast will require much perseverance or some clever electronic rigging to bypass the device.

BACKS: Most radios pictured in this book had backs when they were originally sold. Most don't now. The backs were typically flimsy cardboard which became worn and fell off never to be seen again. Often times the antenna was attached to the missing back and is now also missing. To most collectors the absence of a cardboard back will not detract from the value of a radio. Conversely, the presence of a back, particularly if it is in good condition, will add value. One exception to this rule are those radios made during the mid to late 1930s with molded plastic backs (like the Fada 260 series or the Emerson 199). With these radios the back is considered to be a critical element of the radio design and, as such, its absence can devalue a set by as much as 25%.

KNOBS: If one or more knobs are missing from a radio its value will be diminished. The purchaser of such a radio must resolve himself to either completing the set or substituting another complete set that looks aesthetically correct. If you chose to complete the set with correct knobs, the most cost effective, but also the most time consuming, method is to bring a correct knob with you to a local collector meet in hopes of finding a match in a box of odd knobs for sale. Cost for such knobs is typically in the two to five dollar range. Be prepared to wait a long time and look through a lot of boxes though. Another way to complete the set is to purchase a reproduction from one of several vendors throughout the country. Check ads in trade periodicals and club newsletters for such resources. There are dozens of types being reproduced and stocked today. The price for such reproduction knobs will range from four to twenty dollars. Specialty knobs with chrome rings or those that duplicate catalin can run upwards of twenty-five dollars each. If you've got an odd bird which is not being reproduced, your final alternative is a specialty vendor who will make a casting of your knob with reproductions upon order. Cost for such knobs typically begins at twelve dollars. Regardless of how you decide to deal with the missing knob(s), the value of the radio will be reduced by an amount which correlates directly with the time, cost and aggravation of completing the set.

DIAL LENSES: Covering the dial area on most radios pictured in this book is a plastic or glass dial lens. A damaged or missing plastic dial lens can be reproduced to order by a number of vendors. The going rate is about eighteen dollars per lens. If you don't have the original you can still make a tracing of the dial opening and send it to with your order. Again, look in periodicals and club bulletins for resources.

Glass dial lenses are typically made from convex round pieces of glass secured either inside the dial bezel or attached to the dial scale on the radio itself. Clockmakers usually will have matches for such pieces in the form of clock crystals. Just measure the outside diameter of the glass and make a few calls to area clockmakers or contact a supply house. You might, as an alternative, decide to replace the glass with plastic and contact a lens reproducer.

Both glass and plastic dial lenses were sometimes reverse painted with dial scales and ornamentation. If such a dial is broken, cracked, warped or otherwise damaged you've got a significant problem facing you. Except for a limited number of high value radios, no one makes glass reproduction dial scales. As an alternative, you may chose to replace the glass scale with a reproduction plastic scale. Rock-Sea Enterprises, listed in the resource section of this book, is the only dedicated dial reproduction resource of which I am aware. They produce an excellent product.

PLASTICS

For the novice collector the number of different plastics used in the manufacture of radio cabinets can be confusing. Furthermore, the names we use today have been established more from convention than anything else. That is, most all of the descriptive names currently used were at one time a brand or product name for the material. Typically, many manufacturers produced the same or a similar material. For one reason or another a single brand name has, over time, become synonymous with the material itself. The most common types of plastics and their conventions are as follows:

BAKELITE: The 'first' plastic, Bakelite was made from a powder and molded under tremendous pressure and heat. It is typically brown or black and somewhat porous. Brown bakelite can be heavily marbleized with various shades of coloration.

BEETLE: Formed, like bakelite, under heat and pressure, this early plastic can be striking in appearance. Used to mold cabinets from the late 1930s through the early 1940s, beetle is typically opaque ivory in color with marbled streaks of orange, rust, green, blue, red and brown. Some examples may be subtle with just a hint of rust marbling while others evidence themselves in a wide array of deep color tones.

CATALIN: An early resin, catalin was poured into molds and then cured in low heat ovens over a period of days. Once removed from the mold, if not broken in the process, the catalin radio cabinet was machined to remove debris and to add detail to the design. The labor intensity of this material was not at issue in the late 1930s when it was first introduced in radio cabinetry, but after World War II, with a shortage of manpower, its use quickly became cost prohibitive.

Catalin is known to be the most valuable of plastics used in radios with coloration ranging from opaque solid colors to semi-translucent marbles resembling polished agate. Catalin is more fragile than other plastics of the period and is prone to chips, cracks, burns (from tube heat), discoloration (those butterscotch radios were originally white) and shrinking. All of these hazards have had a significant impact on the number of catalin sets which have survived, thus driving up their price.

PLASKON: Manufactured using a process similar to bakelite, plaskon was molded in ivory and opaque colors. Typically, radio manufacturers offered the consumer the option of an ivory color at a higher price. Sometimes this was simply brown or black bakelite cabinet which had been painted ivory. Other times it was molded in ivory plaskon. Occasionally, radios were offered in colors other than ivory, usually in limited quantities. Plaskon cabinets were available in colors such as pistachio green, Chinese red, sky blue and lavender. Although plaskon wears about as well as bakelite, it is subject to stress lines. These are superficial cracks which, unfortunately, fill with dirt and grime over the years. Some can be cleaned or bleached out, others can be sanded down, but most must be lived with.

PLASTIC: Refers to contemporary injection molded plastics, most frequently polystyrene. These more contemporary plastics were first widely used in the late 1940s and available in a variety of colors from a bakelite brown color to bright colors of the rainbow and in both solid and marbled varieties.

TENITE: Occasionally used for cabinets, tenite was most often used for the manufacture of radio grilles, knobs, handles and ornamental parts. The most significant problem with this material was its vulnerability to shrinking and to warpage from proximity to heat. Tenite was not a particularly wise choice for use in construction of heated filament tube-type radios.   Almost all tenite warps or shrinks to one degree or another. The best one can hope for is minimal warpage.

VALUING PLASTIC RADIOS

CHIPS AND CRACKS: As a rule of thumb, a major flaw in any plastic radio, such as a significant visible chip, crack or warp, will cut the value of that radio in half so long as it remains displayable. Among plaskon radios hairline stress cracks are fairly common. In many models they are the rule and not the exception. As stated previously, so long as the stress lines are not too numerous and do not detract materially from the general aesthetics of the radio, the depreciation should be minimal. On the other hand, the stress free example of a set which is commonly found with stress lines is worth considerably more than average.

FRAGILITY: Dependent on the thickness of the casting, materials used in construction and the extremity of design, some radios are inherently more delicate than others. The most fragile (such as the Kadette 'Classic') are rarely found in near perfect condition. This 'universe' from which each radio is drawn must be considered in determining its value. In our example, a Kadette 'Classic' in what would appear to the casual observer as marginal condition, might, in fact, be an excellent example of that model, given the condition of other surviving sets.

REPAIRS: Although there have been some relatively successful attempts at bakelite repair, no repair can go undetected. A well repaired flaw can increase the value of a radio but will never raise its value to that of an unflawed one.

PAINT CHIPS: While many early plastic radios were available in different colors (typically walnut, black and ivory), oftentimes the ivory colored sets were brown or black bakelite which had been painted at the factory. This factory painting process was similar to that used in the automobile industry. After several layers of paint were applied, the cabinet was baked for several hours. As a result, the paint became 'hard' and much more susceptible to chipping over time. In addition to the problem of chipping, the baking process made the paint that much more difficult to remove. Stripping old baked on paint is extremely difficult and time intensive at best. Be prepared to use caustic chemicals and spend a lot of time. You may then find yourself with a black or brown radio and cast ivory plaskon knobs. This may look fine to you, but it is not 'original' and will be less valuable to other collectors. The net result is that the value of a factory painted radio with paint chips will be reduced by anywhere from ten to fifty percent dependent on the extent and placement of chipping.

VALUING WOOD RADIOS

Although typically not as popular with collectors in the past, wood cabinet radios have recently gained in value and popularity as the hobby has expanded. It is interesting to note that, when radios were sold in the 1930s and 1940s, it was the wood cabinet which sold for a premium. Plastics were considered then to be a cheap man-made alternative to wood.

The following are some basics to keep in mind when considering the purchase of a wood cabinet radio:

FINISH: Most wood radios produced during the 1930s through the 1950s were finished with either clear or toned lacquer. The more expensive sets were hand rubbed resulting in the 'piano' finish many collectors today find so desirable. Over the years many things can and usually do happen to a lacquer finish. The finish will dry and chip, peel or separate ('alligator'). Rarely does one come across an original lacquer finish without some evidence of the passage of time. Unless the surviving finish is horrendous, it is worthwhile trying to salvage. There are various products available which will allow you to easily clean the surface and replenish the moisture in the lacquer. Additionally, you may want to amalgamize the finish to cover bare spots. This involves dissolving the original finish with denatured alcohol or other substance and respreading it.

As opposed to clear lacquer, toned lacquers present more of a problem. Many manufacturers, instead of using different wood veneers for contrast, used a single veneer type and sprayed toned lacquers for variations in color and shade. Originally, the difference was difficult to discern. Today it is evident. When toned lacquers age they reveal the original color of the wood underneath, usually in deep contrast. The darker the toner, the more significant the problem. The purchaser of such a radio must resign himself to live with the radio as is or refinish it. Both are a compromise.

VENEERS: Most all early radio cabinets were made with wood veneers as opposed to solid exotic or hardwoods. Over time the glue which bonds the veneer to the wood of the cabinet can deteriorate. The result ranges from small veneer chips to full pieces of veneer falling off. The best case is where the veneer is separating, but still present. This is easy to remedy with glue and clamps. If pieces of veneer are actually missing, they must obviously be replaced. This means removing the remainder of the damaged section of veneer and replacing it with another full piece. You may be able to find such a piece with an original finish on a junker radio. Otherwise, you will need to apply a finish to the new veneer. If the piece is very small and in a relatively concealed area you may chose to fill the hole with a matching wood putty, shellac stick or toner pencil.

Additionally, veneer tends to lose its flexibility over time. Veneer which has been bent at angles of ninety degrees or more on a radio may break at those turns. If this occurs, again, be prepared to live with it or replace that entire section of veneer.

PAPER VENEER: A concept similar to the toned lacquer, paper veneer involved application of exotic veneer 'decals' to the radio cabinet. This technique was used for both whole radio cabinets and small details. Upon close inspection of the veneer, one can usually discriminate between paper and real veneers. A damaged decal will make the difference apparent. If a radio is ornamented with damaged paper finish, any attempt to repair or refinish the radio will likely remove the decal or damage it further. This should be kept in mind when evaluating such a set for purchase.

REFINISHING: All but the absolute purists will agree that some radios just need to be refinished in order to be displayable. Many collectors have not seen a well refinished radio and so cannot appreciate the art to it. The process is long and involved, and extremely time intensive, but the result of a 'professional' refinish is astonishing. Of course there is refinishing and there is refinishing. Many collectors will use a caustic solvent to remove the original finish, then stain the cabinet and apply a coat of polyurethane or tung oil. The result is something that looks like it came out of a craft shop, not a vintage radio. If you are considering the purchase of a radio which is so refinished it should be valued as if it had no finish at all (about half the value of such a set in average condition).

Other types of refinishing are less objectionable and easier to remedy. Oftentimes one encounters a radio with the original finish intact. It is just underneath a coat of slopped on varnish or shellac. Various solvents can be used to remove the top coat and leave the original finish. There is an art to this process in both identifying the right solvent and its application. Read up on refinishing and talk to other collectors about their experiences before attempting this on a valuable acquisition. The net result on the value of radios so 'refinished' is a reduction by about twenty-five percent. A note of caution: be certain that the slopped on coat is not polyurethane. If it is, you're back to square one and must remove all finishes.

RESOURCES

As the hobby of vintage radio collecting has evolved and expanded over time so have the resources for collectors. There are regional, national and international clubs. Cottage industries have sprung up offering a wide range of products and services. There are club newsletters, bulletins, references books and a monthly periodicals on the subject.

VINTAGE RADIO CLUBS:

By far the best part of the hobby is meeting and spending time with other collectors. Club meets provide a forum for exchange of information, purchase of resources and new radio acquisitions among other activites. There are few collectors in the United States who are not within a few hours drive from a local or regional club. A quick search on the internet can locate your nearest radio collectors' club.

THE INTERNET:

With the recent advent of the Worldwide Web, shopping and resourcing for the hobbyist is both easier and more convenient. With little knowledge and any search engine, one can locate hundreds of radio related websites and find resources for club meetings, supplies, restoration techniques, or just someone to converse with about the hobby. A great place to start would be one of the websites listed in the acknowledgment page of this book. Take a look at the site and the link page to locate other endorsed sites.

In addition to hobby-specific websites, there are now several on-line auction houses, the largest of which are sponsored by Ebay and Amazon.com. As with the antique business in general, on-line auctions have had a significant impact on where vintage radios are bought and sold. Every collector with access to the internet should spend some time browsing through the auction sites. Incredible finds turn up on a regular basis and you can search on your own time schedule.

OTHER RESOURCES

There are a host of suppliers and vendors who will make your life as a collector a little easier. Listed below are a representative few:

GENERAL ELECTRONICS SUPPLIES/REPAIRS:

ANTIQUE ELECTRONIC SUPPLY, 6221 S, MAPLE AVENUE, TEMPE, AZ 85283
WHOLESALE SUPPLIER OF VINTAGE TUBES, ELECTRONIC COMPONENTS, REPRODUCTION KNOBS, GRILLE CLOTH AND OTHER ITEMS.

SPARKY'S ANTIQUE RADIO REPAIR, WESTMINSTER, MD (410) 848-5279
EXPERT GENERAL VINTAGE RADIO ELECTRONIC REPAIR

GENERAL REFINISHING/WOODWORKING:

CONSTANTINE'S, 2050 EASTCHESTER RD., BRONX, NY 10461 (718) 792-1600
COMPLETE LINE OF WOODWORKING TOOLS AND SUPPLIES INCLUDING VENEERS.

KOTTON KLEANSER PRODUCTS, INC., PO BOX 1386, BRADEN, TN 38010-1386
MANUFACTURER OF WOOD CLEANER, PROTECTIVE WOOD FEEDER AND OTHER ANTIQUE RESTORATION PRODUCTS.

MISC. PARTS - VINTAGE AND REPRODUCTION:

GREAT NORTHERN, PO BOX 17338, MINNEAPOLIS, MN 55417
WIDE RANGE OF NOS AND REPRODUCTION ZENITH PARTS AND OTHER ITEMS. WRITE FOR CATALOG.

VINTAGE RADIO & TV SUPPLY, 3498 WEST 105TH STREET, CLEVELAND, OH 44111
REPRODUCTION KNOBS, PARTS, TUBES AND OTHER ITEMS. (216) 671-6712

REPRODUCTION GRILLE CLOTH:

MICHAEL KATZ, 3987 DALEVIEW AVE., SEAFORD, NY 11783
REPRODUCTION RADIO GRILLE CLOTH. SEND $.52 LSASE FOR SAMPLES.

JOHN OKOLOWICZ, 624 CEDAR HILL ROAD, AMBLER, PA 19002
REPRODUCTION RADIO GRILLE CLOTH. (215) 542-1597/EVENINGS, 70313.2564@COMPUSERVE.COM

REPRODUCTION DIAL SCALES AND DECALS:

ROCK-SEA ENTERPRISES, 323 E. MATILIJA ST., #110-241, OJAI, CA 93023
ALL TYPES OF REPRO DIAL SCALES AND DECALS, BOTH CATALOG AND MADE TO ORDER.
E-MAIL: DIALS@JUNO.COM   WEB PAGE: HTTP://MEMBERS.AOL.COM/ROCKSEAENT/

REPRODUCTION KNOBS & SMALL PARTS:

OLD TIME REPLICATIONS, 5744 TOBIAS AVE., VAN NUYS, CA 91411
REPRODUCTION KNOBS AND OTHER PARTS. (818) 909-0241

KRIS GIMMY, THE KNOB MAN, 1441 NOTTINGHAM DR., AIKEN, SC 29801
SPECIALIZING IN CATALIN AND OTHER HIGH-END RADIO KNOBS.

REPRODUCTION PLASTIC DIAL LENSES:

DOYLE ROBERTS, HC63, BOX 236-1, CLINTON, AR 72031
DIAL LENSE REPRODUCTION FROM DAMAGED OLD LENSE OR DRAWING OF DIAL OPENING..

Abbotwares

Large Horse c.1947
metal, 5 tubes, 1 band
$325

Abbotwares

TROTTER c.1947
METAL, 5 TUBES, 1 BAND
$375

ACE

11B c.1937
WOOD, 11 TUBES, 3 BANDS
$400

ACE

47F c.1937
WOOD, 4 TUBES, 1 BAND
$150

ACE

61D c.1937
WOOD, 6 TUBES, 2 BANDS
$225

ACE

71D c.1937
WOOD, 7 TUBES, 2 BANDS
$175

ACE

77B c.1937
WOOD, 7 TUBES, 2 BANDS
$375

ACE

77H c.1937
WOOD, 7 TUBES, 2 BANDS
$225

ACE

91B c.1937
WOOD, 9 TUBES, 3 BANDS
$275

ACE

JUNIOR c.1933
WOOD
$350

ACE

JUNIOR c.1933
WOOD
$350

ACE

MIDGET c.1933
WOOD
$350

AcraTone

4B, "Buddy", c.1936
WOOD, 4 TUBES, 1 BAND
$70

AcraTone

5B, "Dandy", c.1936
WOOD, 4 TUBES, 1 BAND
$60

AcraTone

6B, c.1936
WOOD, 5 TUBES, 2 BANDS
$75

AcraTone

7B, c.1936
WOOD, 5 TUBES, 2 BANDS
$120

AcraTone

7C, c.1936
WOOD, 5 TUBES, 2 BANDS
$125

AcraTone

11B, c.1936
WOOD, 7 TUBES, 3 BANDS
$140

AcraTone

12B, c.1936
WOOD, 7 TUBES, 3 BANDS
$145

AcraTone

13B, c.1936
WOOD, 5 TUBES, 1 BAND
$600

AcraTone

14B, c.1936
WOOD, 6 TUBES, 2 BANDS
$110

AcraTone

23B, c.1936
WOOD, 6 TUBES, 3 BANDS
$175

AcraTone

24B, c.1936
WOOD, 8 TUBES, 3 BANDS
$165

AcraTone

26C, c.1936
WOOD, 7 TUBES, 3 BANDS
$225

AcraTone

29C, C.1936

WOOD, 8 TUBES, 3 BANDS

$225

AcraTone

32C, C. 1936

WOOD, 7 TUBES, 3 BANDS; DC

$125

AcraTone

50, C.1935

WOOD, 5 TUBES, 1 BAND

$80

AcraTone

52, C.1935

WOOD, 5 TUBES, 1 BAND

$65

AcraTone

100, C.1935

WOOD, 4 TUBES, 1 BAND

$300

AcraTone

127, C.1935

WOOD, 4 TUBES, 2 BANDS

$125

AcraTone

141, C.1935

WOOD, 5 TUBES, 1 BAND

$90

AcraTone

142, C.1935

WOOD, 5 TUBES, 2 BANDS

$275

AcraTone

143, C.1935

WOOD, 6 TUBES, 3 BANDS

$175

AcraTone

144, C.1936

WOOD, 5 TUBES, 2 BANDS

$110

AcraTone

146, C.1936

WOOD, 6 TUBES, 2 BANDS

$120

AcraTone

149, BANDSPREAD, C.1935

WOOD, 6 TUBES, 3 BANDS

$750

AcraTone

157, 163, c.1936,
WOOD, 8 TUBES, 8 BANDS
$150

AcraTone

167B, 168B, c.1935
WOOD, 5 TUBES, 2 BANDS; DC
$80

AcraTone

179B, c.1936
WOOD, 6 TUBES, 3 BANDS; DC
$80

AcraTone

192, c/1935
WOOD, 5 TUBES, 2 BANDS
$110

AcraTone

228, c.1935
WOOD, 6 TUBES, 3 BANDS; DC
$80

AcraTone

248, c.1935
WOOD, 6 TUBES, 1 BAND; DC
$90

Admiral.

4R11, c.1951
PLASTIC, 4 TUBES, 1 BAND
$80

Admiral.

5B4, c.1958
PLASTIC, 5 TUBES, 1 BAND
$50

Admiral.

5D4, c.1958
PLASTIC, 5 TUBES, 1 BAND
$50

Admiral.

5E31(BLACK), 5E32(BROWN), 5E33
(IVORY), 5E38(GREEN), 5E39(GREY)
c.1954
PLASTIC, 6 TUBES, 1 BAND
$50

Admiral.

5G31(BLACK), 5G32(BROWN),
5G33(IVORY), c.1954
PLASTIC, 6 TUBES, 1 BAND
$65

Admiral.

5J2, c.1951
BAKELITE, 5 TUBES, 1 BAND
$25

Admiral.

5T31, C.1955

BAKELITE, 5 TUBES, 1 BAND

$35

Admiral.

5X13, C.1951

BAKELITE

$25

Admiral.

6A22, C.1951

BAKELITE

$25

Admiral.

6J2, RADIO-PHONO, C.1952

BAKELITE, 6 TUBES, 1 BAND

$70

Admiral.

8SI, C.1959

TRANSISTOR, 1 BAND, DC

$55

Admiral.

21A6, C.1939

PLASKON, 6 TUBES, 1 BAND

$125

Admiral.

103-6B, C.1938

WOOD, 6 TUBES, 2 BANDS

$70

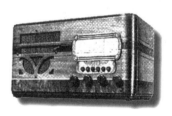

Admiral.

103-6B (ALTERNATE), C.1938

WOOD, 6 TUBES, 2 BANDS

$70

Admiral.

104-4A, C.1940

WOOD, 4 TUBES, 1 BAND, DC

$40

Admiral.

150-5Z(BLACK), 155-5Z(IVORY),
C.1938

BAKELITE/PLASKON, 5 TUBES, 1 BAND

BLACK $175

IVORY $225

Admiral.

159-5L, RADIO-PHONO,
C.1940

WOOD, 5 TUBES, 1 BAND

$75

Admiral.

160-5X, C.1938

WOOD, 5 TUBES, 2 BANDS

$60

Admiral.

165-6W, c.1938
WOOD, 6 TUBES, 2 BANDS
$65

Admiral.

170-7M, c.1938
WOOD, 7 TUBES, 3 BANDS
$65

Admiral.

175-8K, c.1938
WOOD, 8 TUBES, 3 BANDS
$110

Admiral.

205-6Q, c.1938
WOOD, 6 TUBES, 2 BANDS, DC
$60

Admiral.

353-5R, c.1940
WOOD, 5 TUBES, 1 BAND
$125

Admiral.

361-2Q(BLACK), 361-5Q(IVORY),
c.1940
BAKELITE/PLASKON, 5 TUBES, 1 BAND
BLACK $80
IVORY $125

Admiral.

366-6J(BROWN), 367-6J(IVORY), 368-6J(BEETLE)
BAKELITE/PLASKON/BEETLE, 5 TUBES, 1 BAND
BROWN $125
IVORY 225
BEETLE $425

Admiral.

369-6J, c.1940
WOOD, 6 TUBES, 1 BAND
$90

Admiral.

371-5R(BROWN), 372-5R(IVORY), 373-5R(BEETLE)
BAKELITE/PLASKON/BEETLE, 5 TUBES, 1 BAND
BROWN $125
IVORY 225
BEETLE $425

Admiral.

A126, c.1936
WOOD
$140

Admiral.

B225, c.1936
WOOD
$55

Admiral.

CW13, PHONO REMOTE, c.1937
WOOD
$75

Admiral.

L567, C.1937

WOOD, 7 TUBES, 3 BANDS

$175

Admiral.

M351, C.1936

WOOD

$275

Aero

4, C.1933

WOOD, 4 TUBES, 1 BAND

$175

Aero

5, C.1933

WOOD, 5TUBES, 1 BAND

$225

Aero

6, C.1933

WOOD, 6 TUBES, 1 BAND

$250

Aero

SHORT WAVE CONV., C.1931

METAL

$125

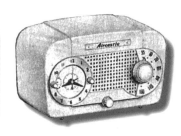

AIR CASTLE

314, C.1951

PLASTIC, 4 TUBES, 1 BAND

$30

AIR CASTLE

315(BROWN), 316(IVORY), 1951

BAKELITE/PLASKON, 5 TUBES, 1 BAND

$20

AIR CASTLE

317(BROWN), 318(IVORY), 1951

BAKELITE/PLASKON, 5 TUBES, 1 BAND

$20

AIR CASTLE

383, C.1951

BAKELITE, 8 TUBES, AM/FM

$100

AIR CASTLE

708, C.1951

PLASTIC, 5 TUBES, 1 BAND

$65

AIR CASTLE

2001, C.1940

WOOD, 8 TUBES, 2 BANDS

$125

AIR CASTLE

2002(BROWN), 2003(IVORY), C.1940
BAKELITE, BED LAMP, 5 TUBES, 1 BAND
$75

AIR CASTLE

2004(BROWN), 2005(IVORY), C.1940
METAL, 3 TUBES, 1 BAND
$150

AIR CASTLE

2006(BROWN), 2007(IVORY), C.1940
BAKELITE/PLASKON, 5 TUBES, 2 BANDS
BROWN $75
IVORY $125

AIR CASTLE

2008 'DIAMOND JUBILEE', C.1940
WOOD, 7 TUBES, 2 BANDS
$1000

AIR CASTLE

2001, C.1940
WOOD, 7 TUBES, 2 BANDS
$90

AIR CASTLE

2012, RADIO-PHONO, C.1940
WOOD, 5 TUBES, 1 BAND
$50

AIR CASTLE

2100, 2200, C.1940
WOOD, 6 TUBES, 2 BANDS, DC
$80

AIR CASTLE

2104, 2106, C.1940
WOOD, 4 TUBES, 1 BAND, DC
$30

AIR CASTLE

2108, 2110, C.1940
WOOD, 5 TUBES, 2 BANDS, DC
$40

AIR CASTLE

4500, C.1935
WOOD, 4 TUBES, 1 BAND
$125

AIR CASTLE

4501, C.1935
WOOD, 5 TUBES, 1 BAND
$75

AIR CASTLE

4502, C.1935
WOOD, 5 TUBES, 2 BANDS
$275

AIR CASTLE
4503, c.1935
WOOD, 6 TUBES, 2 BANDS
$225

AIR CASTLE
4504, c.1935
WOOD, 7 TUBES, 2 BANDS
$110

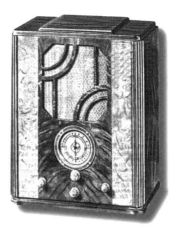

AIR CASTLE
4505, c.1935
WOOD, 9 TUBES, 2 BANDS
$150

AIR CASTLE
4510, c.1935
WOOD, 6 TUBES, 1 BAND, DC
$100

AIR CASTLE
4512, c.1935
WOOD, 7 TUBES, 2 BANDS, DC
$75

AIR CASTLE
5000, c.1937
WOOD, 6 TUBES, 2 BANDS
$115

AIR CASTLE
5002, c.1937
WOOD, 5 TUBES, 2 BANDS
$65

AIR CASTLE
5004, c.1937
WOOD, 7 TUBES, 3 BANDS
$110

AIR CASTLE
5006, c.1937
WOOD, 8 TUBES, 3 BANDS
$110

AIR CASTLE
5008, c.1937
WOOD, 11 TUBES, 3 BANDS
$175

AIR CASTLE
5010, RADIO-PHONO, c.1937
WOOD, 6 TUBES, 2 BANDS
$100

AIR CASTLE
5012, c.1941
WOOD, 7 TUBES, 2 BANDS
$45

5014, c.1941
WOOD, 8 TUBES, 2 BANDS
$90

AIR CASTLE

5012, c.1937
WOOD, 5 TUBES, 2 BANDS, DC
$110

AIR CASTLE

5105, c.1937
WOOD, 7 TUBES, 3 BANDS, DC
$110

AIR CASTLE

5110, c.1941
WOOD, 6 TUBES, 2 BANDS, DC
$80

AIR CASTLE

5202, c.1937
WOOD, 6 TUBES, 3 BANDS, DC
$110

AIR CASTLE

5206, c.1937
WOOD, 8 TUBES, 3 BANDS, DC
$150

AIR CASTLE

6442, c.1937
WOOD, 6 TUBES, 3 BANDS, DC
$75

AIR CASTLE

6500, c.1937
WOOD, 6 TUBES, 3 BANDS
$125

AIR CASTLE

6502, c.1937
WOOD, 7 TUBES, 3 BANDS
$175

AIR CASTLE

6504, c.1937
WOOD, 6 TUBES, 3 BANDS
$200

AIR CASTLE

6506, c.1937
WOOD, 8 TUBES, 4 BANDS
$275

AIR CASTLE

6510, c.1937
WOOD, 5 TUBES, 2 BANDS
$75

AIR CASTLE

6600, C.1937

WOOD, 5 TUBES, 1 BAND, DC

$125

AIR CASTLE

6602, C.1937

WOOD, 7 TUBES, 3 BANDS, DC

$125

AIR CASTLE

6704, C.1937

WOOD, 6 TUBES, 3 BANDS, DC

$75

AIR CASTLE

6708, C.1937

WOOD, 7 TUBES, 4 BANDS, DC

$125

AIR CASTLE

"TULIP", C.1938

WOOD

$500

AIR-KING

C.1935

WOOD, 6 TUBES, 2 BANDS

$175

AIR-KING

335, "ATLAS", C.1935

WOOD, 6 TUBES, 2 BANDS

$750

AIR-KING

456, "SUPER 5", C.1934

WOOD, 5 TUBES, 1 BAND

$185

AIR-KING

459, RADIO-PHONO, C.1947

WOOD, 5 TUBES, 1 BAND

$40

AIR-KING

461, PHONO, C.1947

WOOD, 3 TUBES, NO RADIO

$25

AIR-KING

470, RADIO-PHONO, C.1947

WOOD, 4 TUBES, 1 BAND

$50

AIR-KING

505D, C.1935

WOOD, 5 TUBES, 1 BAND

$125

AIR-KING

507, C.1935
WOOD, 5 TUBES, 1 BAND, DC
$115

AIR-KING

627, SW ADAPTER, C.1932
METAL, 3 TUBES, SHORT WAVE
$125

AIR-KING

2500, C.1935
WOOD, 6 TUBES, 2 BANDS
$325

AIR-KING

"MONARCH", C.1936
WOOD, 5 TUBES, 1 BAND
$70

AIR-KING

"SUPER 7", C.1935
WOOD, 7 TUBES, 1 BAND
$275

AIR-KING

SW CONVERTER, C.1931
METAL, 1 TUBE, SHORT WAVE
$125

AIR-KING

SW CONVERTER, C.1932
METAL, 4 TUBES, SHORT WAVE
$150

AIR-KING

"TRIPLE GRID", C.1935
WOOD, 5 TUBES, 1 BAND
$300

AIRMASTER

910, C.1939
WOOD, 5 TUBES, 1 BAND
$125

AIRMASTER

930, C.1939
WOOD, 4 TUBES, 1 BAND
$375

AIRMASTER

950, C.1939
WOOD, 5 TUBES, 1 BAND
$425

AIRMASTER

960, C.1939
WOOD, 6 TUBES, 1 BAND
$150

Airline

47, C.1933

WOOD, 7 TUBES, 1 BAND

$225

Airline

57, C.1933

WOOD, 5 TUBES, 1 BAND

$85

Airline

81, C.1933

WOOD, 4 TUBES, 1 BAND

$250

Airline

114, C.1935

WOOD, 7 TUBES, 2 BAND, DC

$125

Airline

AIRLINE

131, C.1935

Airline

135, C.1935

WOOD, 6 TUBES, 2 BAND

$175

Airline

138, C.1935

WOOD, 6 TUBES, 1 BAND

$125

Airline

147, C.1935

WOOD, 5 TUBES, 2 BAND

$175

Airline

149, 159, C.1935

WOOD, 5 TUBES, 1 BAND, DC

$125

Airline

169, 171, C.1936

WOOD, 4 TUBES, 1 BAND, DC

$75

Airline

177, C.1936

WOOD, 7 TUBES, 2 BANDS

$175

Airline

185, C.1936

WOOD, 7 TUBES, 2 BANDS

$175

Airline

211, 213, c.1936
WOOD, 5 TUBES, 1 BAND, DC
$75

Airline

217, 219, c.1936
WOOD, 7 TUBES, 2 BANDS, DC
$95

Airline

225, c.1936
WOOD, 5 TUBES, 2 BANDS
$110

Airline

229, c.1936
WOOD, 6 TUBES, 1 BAND, DC
$75

Airline

230, 240, c.1937
WOOD, 5 TUBES, 2 BANDS, DC
$60

Airline

233, c.1936
WOOD, 5 TUBES, 1 BAND
$80

Airline

235, c.1936
WOOD, 5 TUBES, 2 BANDS
$90

Airline

239, MOVIE DIAL, c.1937
WOOD, 7 TUBES, 3 BANDS
$350

Airline

251, 253, c.1937
WOOD, 8 TUBES, 3 BANDS, DC
$110

Airline

254, c.1937
WOOD, 4 TUBES, 1 BAND, DC
$100

Airline

256, c.1937
WOOD, 6 TUBES, 2 BANDS
$110

Airline

264, c.1937
WOOD, 5 TUBES, 1 BAND, DC
$80

Airline

274 (BLACK), 288 (BROWN),
290 (IVORY), C.1938
BAKELITE/PLASKON, 6 TUBES, 1 BAND
BLACK $200, BROWN $175, IVORY
$250

Airline

277, MOVIE DIAL, C.1937
WOOD, 7 TUBES, 2 BANDS
$350

Airline

280, C.1938
BAKELITE, 5 TUBES, 1 BAND, DC
$100

Airline

283, MOVIE DIAL, C.1937
WOOD, 8 TUBES, 3 BANDS
$375

Airline

294, C.1937
WOOD, 7 TUBES, 2 BANDS, DC
$125

Airline

297, C.1937
WOOD, 7 TUBES, 2 BANDS
$175

Airline

315, C.1937
WOOD, 5 TUBES, 2 BANDS
$100

Airline

316, C.1937
WOOD, 6 TUBES, 2 BANDS
$125

Airline

317, C.1937
WOOD, 7 TUBES, 3 BANDS
$175

Airline

318, MOVIE DIAL, C.1937
WOOD, 8 TUBES, 3 BANDS
$375

Airline

320 (IVORY), 325 (BROWN), C.1939
PLASKON/BAKELITE, 5 TUBES, 1 BAND
$110

Airline

322, C.1939
WOOD, 7 TUBES, 2 BANDS, DC
$85

Airline

329, C.1937
WOOD, 6 TUBES, 2 BANDS, DC
$125

Airline

336, 339, C.1937
WOOD, 6 TUBES, 2 BANDS, DC
$65

Airline

337, MOVIE DIAL, C.1937
WOOD, 7 TUBES, 2 BANDS
$350

Airline

338, C.1937
WOOD, 8 TUBES, 3 BANDS, DC
$115

Airline

343, C.1939
WOOD, 7 TUBES, 2 BANDS, DC
$75

Airline

348, C.1937
WOOD, 4 TUBES, 1 BAND, DC
$125

Airline

350 (BLACK), 351 (BROWN),
352 (IVORY), C.1938
BAKELITE/PLASKON, 5 TUBES, 1 BAND
BLACK $225, BROWN $200, IVORY
$275

Airline

354, C.1939
WOOD, 5 TUBES, 2 BANDS, DC
$45

Airline

355 (BLACK), 365 (BROWN),
375 (IVORY), C.1938
BAKELITE/PLASKON, 5 TUBES, 1 BAND
BLACK $225, BROWN $200, IVORY
$275

Airline

361, C.1939
WOOD, 6 TUBES, 1 BAND
$95

Airline

362, RADIO-PHONO, C.1939
WOOD, 6 TUBES, 2 BANDS
$95

Airline

363, C.1939
WOOD, 6 TUBES, 2 BANDS, DC
$45

Airline

370, C.1939
WOOD, 7 TUBES, 2 BANDS
$110

Airline

386(BLACK), 636(BROWN),
646(IVORY), C.1938
BAKELITE/PLASKON, 6 TUBES, 1 BANDS
BLACK $175, BROWN $150, IVORY
$250

Airline

390, C.1939
WOOD, 9 TUBES, 3 BANDS
$125

Airline

404, C.1938
WOOD, 4 TUBES, 1 BAND, DC
$50

Airline

405, 414, C.1938
WOOD, 5 TUBES, 2 BANDS, DC
$45

Airline

406, 'BULLET', C.1938
WOOD, 6 TUBES, 2 BANDS
$250

Airline

415, C.1937
WOOD, 5 TUBES, 2 BANDS
$75

Airline

416, C.1937
WOOD, 6 TUBES, 3 BANDS
$95

Airline

418, C.1937
WOOD, 8 TUBES, 3 BANDS
$175

Airline

419, C.1937
WOOD, 9 TUBES, 3 BANDS
$175

Airline

420(BROWN), 421(IVORY), 423(RED),
424(BLUE), 431(GREEN), C.1940
BAKELITE/PAINTED, 4 TUBES, 1 BAND
$90

Airline

420(BROWN), 421(IVORY), C.1941
BAKELITE/PAINTED
4 TUBES, 1 BAND
$70

Airline

425, 265(IVORY) C.1937
WOOD, 5 TUBES, 1 BANDS
$110

Airline

437, MOVIE DIAL, C.1937
WOOD, 7 TUBES, 3 BANDS
$375

Airline

439, C.1937
WOOD, 6 TUBES, 2 BANDS, DC
$65

Airline

445 (IVORY), 455 (BROWN),
475 (BLACK), C.1938
BAKELITE/PLASKON, 5 TUBES, 1 BAND
BLACK $225, BROWN $200
IVORY $275

Airline

450, CRYSTAL KIT, C.1959
CRYSTAL
$50

Airline

452, ONE-TUBE KIT, C.1959
1 TUBE, 1 BAND
$50

Airline

452, C.1938
WOOD, 7 TUBES, 2 BANDS, DC
$150

Airline

458, C.1938
WOOD, 8 TUBES, 3 BANDS
$175

Airline

459, C.1939
BAKELITE, 4 TUBES, 1 BAND, DC
$75

Airline

465, C.1939
BAKELITE, 5 TUBES, 1 BAND, DC
$75

Airline

472, C.1938
WOOD, 7 TUBES, 2 BANDS
$135

Airline

486, 'BULLET', C.1938
WOOD, 6 TUBES, 2 BAND, DC
$110

34

Airline

508(BROWN), 509(IVORY), C.1940
BAKELITE/PLASKON, 5 TUBES, 1 BAND
BROWN $110, IVORY $175

Airline

510, RADIO-PHONO, C.1940
WOOD, 5 TUBES, 1 BAND
$60

Airline

511(BROWN), 512(IVORY), C.1941
BAKELITE/PAINTED
5 TUBES, 1 BAND
$70

Airline

513(BROWN), 514(IVORY), C.1941
BAKELITE/PAINTED
5 TUBES, 1 BAND
$100

Airline

515, RADIO-PHONE, C.1941
WOOD, 5 TUBES, 1 BAND
$55

Airline

518(BROWN), 519(IVORY), C.1942
BAKELITE/PAINTED
5 TUBES, 1 BAND
$100

Airline

525(BROWN), 526(IVORY), C.1941
BAKELITE/PAINTED
5 TUBES, 1 BAND
$110

Airline

552, C.1939
WOOD, 5 TUBES, 1 BAND, DC
$55

Airline

562, C.1940
BAKELITE, 5 TUBES, 1 BAND
$225

Airline

564, C.1940
WOOD, 5 TUBES, 1 BAND
$75

Airline

570, C.1941
WOOD, 5 TUBES, 1 BAND
$75

Airline

575, C.1942
WOOD, 5 TUBES, 1 BAND
$65

Airline

600, c.1939

WOOD, 6 TUBES, 2 BANDS

$115

Airline

601, c.1939

WOOD, 6 TUBES, 1 BAND

$85

Airline

602(BROWN), 603(IVORY), c.1940

BAKELITE/PAINTED

6 TUBES, 1 BAND

$195

Airline

604(BROWN), 605(IVORY), c.1940

BAKELITE/PAINTED

6 TUBES, 1 BAND

$225

Airline

610(BROWN), 611(IVORY), c.1941

BAKELITE/PAINTED

6 TUBES, 1 BAND

$110

Airline

612, c.1941

WOOD, 6 TUBES, 2 BANDS

$80

Airline

613, FM ADAPTER, c.1942

WOOD, 7 TUBES, FM ONLY

$75

Airline

613, FM ADAPTER, c.1941

WOOD, 7 TUBES, FM ONLY

$85

Airline

615, RADIO-PHONO, c.1941

WOOD, 6 TUBES, 1 BAND

$70

Airline

617, 618, c.1938

WOOD, 7 TUBES, 2 BANDS, DC

$125

Airline

624(BROWN), 625(IVORY), c.1942

BAKELITE/PAINTED

6 TUBES, 1 BAND

$95

Airline

651, c.1939

WOOD, 6 TUBES, 2 BANDS, DC

$45

Airline

666, C.1940
WOOD, 6 TUBES, 2 BANDS, DC
$65

Airline

675, C.1941
WOOD, 6 TUBES, 2 BANDS
$125

Airline

688, C.1942
WOOD, 6 TUBES, 2 BANDS
$70

Airline

702, C.1939
WOOD, 7 TUBES, 2 BANDS
$85

Airline

704, C.1939
WOOD, 7 TUBES, 2 BANDS
$90

Airline

715, C.1940
WOOD, 7 TUBES, 2 BANDS
$95

Airline

717, RADIO-PHONO, C.1939
WOOD, 7 TUBES, 2 BANDS
$100

Airline

720, C.1940
WOOD, 7 TUBES, 2 BANDS
$90

Airline

729, C.1941
WOOD, 7 TUBES, 2 BANDS
$135

Airline

731, C.1941
WOOD, 7 TUBES, 2 BANDS
$75

Airline

734(BROWN), 735(IVORY), C.1942
BAKELITE/PAINTED
7 TUBES, 2 BANDS
$125

Airline

736, C.1942
WOOD, 6 TUBES, 1 BAND
$125

Airline
750, c.1939
WOOD, 7 TUBES, 2 BANDS
$100

Airline
752, c.1939
WOOD, 7 TUBES, 2 BANDS, DC
$60

Airline
800, c.1940
WOOD, 8 TUBES, 2 BANDS
$95

Airline
803, c.1940
WOOD, 8 TUBES, 5 BANDS
$175

Airline
806, c.1942
WOOD, 8 TUBES, 2 BANDS
$150

Airline
913, c.1942
WOOD, 9 TUBES, 4 BANDS
$125

Airline
917, PHONO ONLY, c.1947
BAKELITE, 2 TUBES
$60

Airline
1064, c.1950
METAL & PLASTIC,
5 TUBES, 1 BAND, DC
$65

Airline
1160, c.1950
5 TUBES, 1 BAND
$40

Airline
1165, c.1950
6 TUBES, 1 BAND
$40

Airline
1405(BROWN), 1406(IVORY), c.1948
BAKELITE/PAINTED
4 TUBES, 1 BAND, DC
$55

Airline
1408, c.1948
WOOD, 4 TUBES, 1 BAND, DC
$30

Airline

1409, c.1948
WOOD, 5 TUBES, 1 BAND, DC
$30

Airline

1411, c.1950
WOOD, 5 TUBES, 1 BAND
$50

Airline

1420, c.1932
WOOD, 4 TUBES, 1 BAND
$175

Airline

1453(BROWN), 1454(IVORY), c.1939
BAKELITE/PAINTED
4 TUBES, 1 BAND
$145

Airline

1460, c.1940
WOOD, 4 TUBES, 1 BAND, DC
$50

Airline

1462, c.1940
WOOD, 4 TUBES, 1 BAND, DC
$150

Airline

1468, c.1941
WOOD, 4 TUBES, 1 BAND
$60

Airline

1469, c.1942
WOOD, 4 TUBES, 1 BAND, DC
$35

Airline

1501(BROWN), 1502(IVORY), c.1948
BAKELITE/PAINTED
5 TUBES, 1 BAND
$75

Airline

1503(BROWN), 1504(IVORY), c.1948
BAKELITE/PAINTED
5 TUBES, 1 BAND
$50

Airline

1507(BROWN), 1508(IVORY), c.1939
BAKELITE/PAINTED
5 TUBES, 1 BAND
$225

Airline

1509(BROWN), 1510(IVORY), c.1948
BAKELITE/PAINTED
5 TUBES, 1 BAND
$50

Airline

1513(BROWN), 1514(IVORY), C.1948
BAKELITE/PAINTED
6 TUBES, 1 BAND
$45

Airline

1520(IVORY), 1521(RED),
1524(BLUE), C.1942
BAKELITE/PAINTED
4 TUBES, 1 BAND
$55

Airline

1525(BROWN), 1526(IVORY), C.1950
BAKELITE/PAINTED
5 TUBES, 1 BAND
$50

Airline

1527(BROWN), 1528(IVORY), C.1950
BAKELITE/PAINTED
5 TUBES, 1 BAND
$40

Airline

1529(BROWN), 1530(IVORY), C.1950
BAKELITE/PAINTED
7 TUBES, AM/FM
$50

Airline

1535, C.1950
BAKELITE, 7 TUBES, AM/FM
$75

Airline

1536(BROWN), 1537(IVORY), C.1950
BAKELITE/PAINTED
5 TUBES, 1 BAND
$60

Airline

1542, C.1955
BAKELITE, 5 TUBES, 1 BAND
$50

Airline

1543(BROWN), 1544(IVORY), C.1952
BAKELITE/PAINTED
5 TUBES, 1 BAND
$25

Airline

1547, C.1952
BAKELITE, 5 TUBES, 1 BAND
$50

Airline

1548(BROWN), 1549(IVORY), C.1955
BAKELITE/PAINTED
5 TUBES, 1 BAND
$40

Airline

1553(BROWN), 1554(IVORY), C.1952
BAKELITE/PAINTED
5 TUBES, 1 BAND
$40

Airline

1553(BROWN), 1554(IVORY), C.1952
BAKELITE/PAINTED
5 TUBES, 1 BAND
$40

Airline

1557(BROWN), 1558(IVORY),
1559(GREEN), C.1955
BAKELITE, 5 TUBES, 1 BAND
$45

Airline

1563, 2663, C.1940
WOOD, 5 TUBES, 2 BANDS, DC
$35

Airline

1564(IVORY), 1565(BROWN),
1566 (RED), 1567(GREEN), C.1955
BAKELITE, 4 TUBES, 1 BAND
$35

Airline

1568, 2568, C.1941
WOOD, 5 TUBES, 2 BANDS
$85

Airline

1572, C.1955
BAKELITE, 8 TUBES, AM/FM
$45

Airline

1576, C.1958
BAKELITE, 4 TUBES, 1 BAND
$55

Airline

1577(BROWN), 1578(IVORY), C.1955
BAKELITE/PAINTED
4 TUBES, 1 BAND
$40

Airline

1579(BROWN), 1580(IVORY), C.1950
BAKELITE/PAINTED, 5 TUBES, 3 BANDS
$75

Airline

1590(BROWN), 1591(IVORY),
1592(RED), C.1950
BAKELITE/PAINTED, 5 TUBES, 1 BAND
BROWN/IVORY $50, RED $75

Airline

1611, C.1932
WOOD, 7 TUBES, 1 BAND
$250

Airline

1615, C.1958
BAKELITE, 4 TUBES, 1 BAND
$20

Airline

1628, C.1959

PLASTIC, 4 TUBES, 1 BAND

$25

Airline

1637, C.1958

BLACK PLASTIC, 6 TUBES, 1 BAND

$25

Airline

1638, C.1959

GREY PLASTIC, 6 TUBES, 1 BAND

$45

Airline

1639, C.1960

TAN & IVORY PLASTIC, 6 TUBES, 1 BAND

$50

Airline

1645(GREEN), 1646 (BROWN),
C.1958

PLASTIC, 6 TUBES, 1 BAND

GREEN $50, BROWN $25

Airline

1653(IVORY), 1654 (BLUE),
C.1958

PLASTIC, 6 TUBES, 1 BAND

IVORY $35, BROWN $65

Airline

1654, 2654, C.1939

WOOD, 6 TUBES, 2 BANDS, DC

$30

Airline

1655(BROWN), 1656(IVORY),
1658(PINK), C.1958

PLASTIC, 4 TUBES, 1 BAND

BROWN/IVORY $30, PINK $50

Airline

1659, 2659, C.1940

WOOD, 6 TUBES, 2 BANDS, DC

$30

Airline

1660(IVORY), 1661(BLUE),
1665(GREY), C.1959

PLASTIC, 5 TUBES, 1 BAND

IVORY/GREY $30, BLUE $50

Airline

1662(IVORY), 1663(BLUE), C.1958

PLASTIC, 5 TUBES, 1 BAND

IVORY $20, BLUE $40

Airline

1666, C.1960

BROWN & IVORY PLASTIC, 8 TUBES,
AMFM

$50

Airline

1667(PINK & IVORY),
1668(RED & IVORY), C.1960
PLASTIC, 5 TUBES, 1 BAND
$40

Airline

1670, C.1959
BLACK PLASTIC
5 TUBES, 1 BAND
$25

Airline

1672(BLUE), 1676(IVORY), C.1960
PLASTIC, 5 TUBES, 1 BAND
BLUE $45, IVORY $25

Airline

1673(BROWN), 1674(BLUE), C.1959
PLASTIC, 5 TUBES, 1 BAND
BLUE $65, BROWN $35

Airline

1801, C.1947
WOOD, 5 TUBES, 1 BAND
$35

Airline

1802, C.1948
WOOD, 5 TUBES, 1 BAND
$35

Airline

1804, C.1948
WOOD, 6 TUBES, 1 BAND
$30

Airline

1811, C.1950
WOOD, 8 TUBES, AMFM
$35

Airline

1813, C.1953
WOOD, 8 TUBES, AMFM
$30

Airline

1814(WALNUT), 1820(OAK), C.1955
WOOD, 6 TUBES, 1 BAND
$30

Airline

1850(PINK), 1851(BLUE), C.1960
PLASTIC, 5 TUBES, 1 BAND
$40

Airline

2002, RADIO-PHONO, C.1947
WOOD, 6 TUBES, 1 BAND
$40

Airline

2007, RADIO-PHONO, C.1947
WOOD, 5 TUBES, 1 BAND
$40

Airline

2030, C.1960
PLASTIC, TRANSISTORIZED
$50

Airline

2074, PHONO REMOTE, C.1940
METAL, PHONO ONLY
$45

Airline

2075, PHONO REMOTE, C.1940
BAKELITE, PHONO ONLY
$45

Airline

2126, C.1932
WOOD, 5 TUBES, 1 BAND
$225

Airline

2557, C.1939
WOOD, 5 TUBES, 1 BAND, DC
$25

Airline

2574, C.1942
WOOD, 5 TUBES, 2 BANDS, DC
$30

Airline

2576, C.1942
WOOD, 5 TUBES, 1 BAND, DC
$30

Airline

2673, C.1941
WOOD, 6 TUBES, 2 BANDS
$110

Airline

2685, C.1942
WOOD, 6 TUBES, 2 BANDS
$110

Airline

3833, PHONO REMOTE, C.1940
BAKELITE, PHONO ONLY
$45

Airline

PREMIUM LAMP, C.1938
CERAMIC
$250

AIR ROAMER

SW1, c.1936

METAL, 1 TUBE, SHORT WAVE

$75

AIR ROAMER

"1-TUBE", c.1936

METAL, 1 TUBE, 1 BAND

$75

AIR ROAMER

"2-TUBE", c.1936

METAL, 2 TUBES, 1 BAND

$100

AIR ROAMER

"3-TUBE", c.1936

METAL, 3 TUBES, 1 BAND

$110

ALL·AMERICAN

PAC (SIMPLEX), c.1935

WOOD, 5 TUBES, 3 BANDS

$150

ALL·AMERICAN

V (SIMPLEX), c.1935

WOOD, 5 TUBES, 2 BANDS

$100

ALL·AMERICAN

W (SIMPLEX), c.1935

WOOD, 8 TUBES, 5 BANDS

$325

ALL·AMERICAN

X (SIMPLEX), c.1935

WOOD, 5 TUBES, 1 BAND

$200

ALL·AMERICAN

Y (SIMPLEX), c.1935

WOOD, 5 TUBES, 2 BANDS

$225

ALL-STAR

c.1938

WOOD, 5 TUBES, 1 BAND

$60

ALL-STAR

"JUNIOR"(KIT), c.1935

METAL, 5 TUBES, 2 BANDS

$125

American

C320-16, c.1937

WOOD, 5 TUBES, 3 BANDS

$75

American

C321-811, c.1937
WOOD, 5 TUBES, 3 BANDS
$150

American

C4185, c.1937
WOOD, 5 TUBES, 1 BAND
$75

American

C4187, c.1937
WOOD, 5 TUBES, 2 BANDS
$125

American

C4215, c.1937
WOOD, 5 TUBES, 1 BAND
$75

American

C6141, c.1937
WOOD, 6 TUBES, 2 BANDS
$125

American

C6188, c.1937
WOOD, 6 TUBES, 1 BAND
$135

American

C6321-61, c.1937
WOOD, 6 TUBES, 3 BANDS
$115

American

C6321-811, c.1937
WOOD, 6 TUBES, 3 BANDS, DC
$100

American

C6322X-46, c.1937
WOOD, 7 TUBES, 3 BANDS
$375

American

C6322XE-29, c.1937
WOOD, 8 TUBES, 3 BANDS
$150

American

C6322XE-411, c.1937
WOOD, 8 TUBES, 3 BANDS
$275

American

C7371E-62, c.1937
WOOD, 8 TUBES, 3 BANDS
$175

American

"MIDGET 5", C.1932
WOOD, 5 TUBES, 1 BAND
$200

American

"MIDGET 6", C.1932
WOOD, 6 TUBES, 1 BAND
$225

American

"MIDGET", C.1931
WOOD, 6 TUBES, 1 BAND
$200

American

RADIO-PHONO, C.1932
WOOD, 6 TUBES, 1 BAND
$375

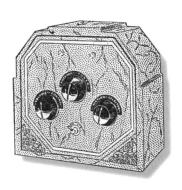

American

SHORT WAVE CONVERTER, C.1932
3 TUBES, SHORT WAVE
$125

AMERICAN-BOSCH

226K, "AIR KING", C.1932
WOOD, 8 TUBES, DC
$150

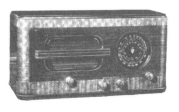

AMERICAN-BOSCH

640, C.1935
6 TUBES, 2 BANDS
$60

AMERICAN-BOSCH

"SUPER 5", C.1933
WOOD, 5 TUBES, 1 BAND
$80

Arkay

421, C.1935
WOOD, 4 TUBES, 1 BAND
$200

Arkay

A, "MARVEL", C.1935
WOOD, 4 TUBES, 1 BAND
$375

Arlington

731, C.1935
WOOD, 5 TUBES, 1 BAND
$125

Arlington

"EXTRA", C.1936
WOOD, 5 TUBES, 1 BAND
$75

Arlington

'PHANTOM RAY', C.1936
WOOD, 6 TUBES, 1 BAND
$100

ARVIN

58, C.1939
BAKELITE
$90

ARVIN

68, C.1939
BAKELITE
$110

ARVIN

78, C.1939
WOOD
$75

ARVIN

88, RADIO-PHONO, C.1940
WOOD, 5 TUBES, 1 BAND
$50

ARVIN

89, C.1941
WOOD, 6 TUBES, 1 BAND
$90

ARVIN

422 (BROWN), 422A (IVORY), C.1941
METAL, 4 TUBES, 1 BAND
$75

ARVIN

502, C.1940
METAL, 5 TUBES, 1 BAND
$175

ARVIN

522 (BROWN), 522A (IVORY), C.1941
METAL, 5 TUBES, 1 BAND
$110

ARVIN

722, C.1941
BAKELITE, 6 TUBES, 1 BAND
$60

Atlantic

'DELUXE 7', C.1932
WOOD, 7 TUBES, 1 BAND
$350

Atlantic

'MIDGET 5', C.1933
WOOD, 5 TUBES, 1 BAND
$200

Atlantic

'MIDGET', C.1933
WOOD, 4 TUBES, 1 BAND
$225

ATWATER KENT

337, C.1935
WOOD, 7 TUBES, 2 BANDS
$375

ATWATER KENT

456, C.1936
WOOD, 6 TUBES, 2 BANDS
$325

ATWATER KENT

637, C.1936
WOOD, 7 TUBES, 2 BANDS
$375

ATWATER KENT

725, C.1936
WOOD, 5 TUBES, 1 BAND
$175

ATWATER KENT

856, C.1936
WOOD, 6 TUBES, 2 BANDS
$350

AUDIOLA RADIO

416, C.1932
PRESSED BOARD, 4 TUBES, 1 BAND
$250

AUDIOLA RADIO

506, C.1932
WOOD, 5 TUBES, 1 BAND
$250

AUDIOLA RADIO

506, C.1934
WOOD, 5 TUBES, 2 BANDS, DC
$150

AUDIOLA RADIO

516, C.1934
WOOD, 5 TUBES, 1 BANDS, DC
$175

AUDIOLA RADIO

710, C.1932
WOOD, 7 TUBES, 1 BAND
$325

AUDIOLA RADIO

1816, C.1934
WOOD, 8 TUBES, 1 BAND, DC
$175

Autocrat

4Z, c.1936

WOOD, 4 TUBES, 1 BAND

$65

Autocrat

5M, c.1937

WOOD, 5 TUBES, 1 BAND

$225

Autocrat

5C, c.1936

WOOD, 5 TUBES, 2 BANDS

$90

Autocrat

5D, c.1936

WOOD, 5 TUBES, 2 BANDS

$125

Autocrat

6G, 6GM, c.1936

WOOD, 6 TUBES, 3 BANDS

$150

Autocrat

8J, 8JM, c.1936

WOOD, 8 TUBES, 5 BANDS

$175

Autocrat

'Junior', c.1933

WOOD, 4 TUBES, 1 BAND

$225

Automatic

8-15, c.1937

BAKELITE/PLASKON, 5 TUBES, 1 BAND

BLACK $275, IVORY $400,

RED $1000, GREEN $1200

Automatic

'BEDLAMP', c.1940

WOOD, 5 TUBES, 1 BAND

$275

BAIRD

SHORTWAVE CONVERTER, c.1932

WOOD, 3 TUBES, SHORT WAVE

$150

BAIRD

SHORTWAVE RECEIVER,

c.1931

WOOD, KIT, 8 TUBES, SHORTWAVE

$75

BELMONT

19, c.1947

BAKELITE, 4 TUBES, 1 BAND, DC

$75

BELMONT
110, c.1947
BAKELITE, 6 TUBES, 1 BAND
$100

BELMONT
115, c.1946
BAKELITE, 4 TUBES, 1 BAND, DC
$350

BELMONT
116, c.1947
BAKELITE, 5 TUBES, 1 BAND
$175

BELMONT
117, c.1947
BAKELITE, 6 TUBES, 1 BAND
$100

BELMONT
118, c.1938
BAKELITE, 5 TUBES, 1 BAND
$250

BELMONT
121, c.1946
BAKELITE, 6 TUBES, 1 BAND
$175

BELMONT
460C, c.1941
WOOD, 4 TUBES, 1 BAND, DC
$50

BELMONT
509, c.1941
WOOD, 5 TUBES, 5 BANDS
$125

BELMONT
513, c.1941
WOOD, 5 TUBES, 1 BAND
$75

BELMONT
518, c.1941
BAKELITE, 5 TUBES, 1 BAND
$200

BELMONT
522, RADIO-PHONO, c.1941
WOOD, 5 TUBES, 1 BAND
$75

BELMONT
534, c.1941
BAKELITE, 5 TUBES, 1 BAND
$225

BELMONT

551, C.1941

WOOD, 5 TUBES, 1 BAND, DC

$75

BELMONT

553, C.1941

WOOD, 5 TUBES, 1 BAND, DC

$75

BELMONT

571, C.1941

WOOD, 5 TUBES, 1 BAND

$175

BELMONT

575, C.1935

WOOD, 5 TUBES, 1 BAND

$200

BELMONT

638, C.1941

BAKELITE, 6 TUBES, 1 BAND

$225

BELMONT

675, C.1935

WOOD, 5 TUBES, 1 BAND

$200

BELMONT

695, C.1941

WOOD, 6 TUBES, 2 BANDS

$75

BELMONT

729, C.1941

WOOD, 7 TUBES, 3 BANDS

$100

Blonder-Tongue

T-88, C.1959

PLASTIC

$50

Bottle Radios

C.1951

PLASTIC, 5 TUBES, 1 BAND

Pepsi $500

Others $150-250

Brown

101, C.1935

WOOD, 4 TUBES, 1 BAND, DC

$75

Brown

106, C.1935

WOOD, 5 TUBES, 2 BANDS

$100

Brown

116, c.1935
WOOD, 6 TUBES, 3 BANDS, DC
$100

Brown

161, c.1935
WOOD, 4 TUBES, 1 BAND, DC
$100

Brown

162, c.1935
WOOD, 6 TUBES, 3 BANDS, DC
$100

Brown

171, c.1935
WOOD, 6 TUBES, 2 BANDS, DC
$100

Brown

172, c.1935
WOOD, 6 TUBES, 2 BANDS
$150

Brown

502, c.1939
WOOD, 5 TUBES, 1 BAND, DC
$100

Brown

545, c.1939
BAKELITE, 5 TUBES, 1 BAND, DC
$75

Brown

556(BROWN), 558(IVORY), c.1939
BAKELITE, 5 TUBES, 1 BAND
brown $150, ivory $250

Brown

560, c.1939
WOOD, 5 TUBES, 1 BAND, DC
$65

Brown

565, c.1939
WOOD, 6 TUBES, 2 BANDS, DC
$75

BRUNSWICK

100, c.1932
WOOD, SHORTWAVE CONVERTER, 2 TUBES
$75

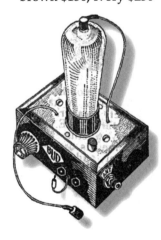

BUD

POLICE THRILLER, c.1933
METAL, 1 TUBE, SHORTWAVE
$100

Camden

5T, c.1938

WOOD, 5 TUBES, 1 BAND

$125

Camden

5TD, c.1938

WOOD, 5 TUBES, 2 BANDS

$125

Camden

6TD(ALT), c.1938

WOOD, 6 TUBES, 3 BANDS

$175

Camden

155-6Y, c.1937

BAKELITE, 6 TUBES, 1 BAND

$175

Camden

160-5X, c.1938

WOOD, 5 TUBES, 2 BAND

$75

Camden

165-6W, c.1938

WOOD, 6 TUBES, 2 BANDS

$100

Camden

170-6P, c.1938

WOOD, 6 TUBES, 3 BANDS, DC

$75

Camden

170-7M, c.1938

WOOD, 7 TUBES, 3 BANDS

$100

Camden

175-8K, c.1938

WOOD, 8 TUBES, 3 BANDS

$175

Camden

205-6Q, c.1938

WOOD, 6 TUBES, 2 BANDS, DC

$75

Camden

376, c.1937

WOOD, 6 TUBES, 3 BANDS

$150

Camden

376L, c.1937

WOOD, 6 TUBES, 3 BANDS

$150

Camden

376U, c.1937

WOOD, 6 TUBES, 3 BANDS

$150

Camden

377L, c.1937

WOOD, 7 TUBES, 3 BANDS

$150

Camden

377U, c.1937

WOOD, 7 TUBES, 3 BANDS

$150

Camden

378, c.1937

WOOD, 8 TUBES, 3 BANDS

$175

Camden

439, c.1939

WOOD, 4 TUBES, 1 BAND

$200

Camden

439MR, c.1939

WOOD, 4 TUBES, 1 BAND

$200

Camden

512-6D, c.1939

WOOD, 6 TUBES, 2 BANDS, DC

$50

Camden

515-5A, c.1939

BAKELITE, 5 TUBES, 1 BAND

$225

Camden

524, "TINY TIM", c.1939

BAKELITE, 4 TUBES, 1 BAND

$90

Camden

528, c.1938

WOOD, 5 TUBES, 1 BAND

$110

Camden

522-6C, c.1939

WOOD, 6 TUBES, 2 BANDS. DC

$75

Camden

538-8A, C.1939

WOOD, 8 TUBES, 3 BANDS

$125

Camden

539-MR, c.1939

WOOD, 5 TUBES, 1 BAND

$200

Camden

541-4A, c.1939

WOOD, 4 TUBES, 1 BAND, DC

$35

Camden

633S, c.1938

WOOD, 6 TUBES, 1 BAND

$75

Camden

"DELUXE 7", c.1933

WOOD, 7 TUBES, 1 BAND

$225

Camden

"SUPER 12", c.1937

WOOD, 12 TUBES, 2 BANDS

$175

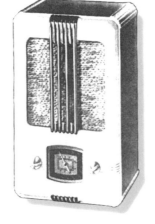

CAMEO

5B1, c.1938

BAKELITE/PLASKON, 5 TUBES, 1 BAND

BROWN $175

IVORY $250

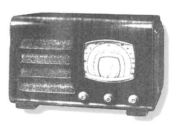

CAMEO

5TA, c.1938

WOOD, 5 TUBES, 2 BANDS

$65

CAMEO

6TA2, c.1938

WOOD, 5 TUBES, 2 BANDS

$90

CAMEO

6TA, c.1938

WOOD, 6 TUBES, 2 BANDS

$85

CAMEO

6TA6, c.1938

WOOD, 6 TUBES, 2 BANDS

$85

CAMEO

7TA, c.1938

WOOD, 7 TUBES, 3 BANDS

$85

CAMEO

8TA, c.1938

WOOD, 8 TUBES, 3 BANDS

$125

Capehart

2T55, c.1956

PLASTIC, 5 TUBES, 1 BAND

$35

Capehart

3T55, c.1956

PLASTIC, 5 TUBES, 1 BAND

$45

Capehart

76C66, c.1956

PLASTIC, 6 TUBES, 1 BAND

$35

Capehart

T54, c.1954

PLASTIC, 5 TUBES, 1 BAND

$45

Cardinal

"MIDGET COMBO", c.1931

WOOD, RADIO-PHONO, 6 TUBES, 1 BAND

$275

Carron

"BULLET", c.1942

BAKELITE, CRYSTAL

$250

Carron

"MIDGET", c.1942

WOOD, CRYSTAL

$150

Cavalier

c.1955

UREA, 5 TUBES, 1 BAND

$75

CENTURY

412, c.1937

WOOD, 4 TUBES, 1 BAND

$65

CGE
Canadian General Electric

C48B, c.1950

BAKELITE, 4 TUBES, 1 BAND

$50

CGE
Canadian General Electric

C65, c.1958

PLASTIC, 6 TUBES, 1 BAND

$30

CGE
Canadian General Electric

C141, c.1958

PLASTIC, 4 TUBES, 1 BAND

$50

CGE
Canadian General Electric
C145, c.1956
PLASTIC, 6 TUBES, 1 BAND
$30

CGE
Canadian General Electric
C154, c.1950
4 TUBES, 1 BAND
$50

CGE
Canadian General Electric
C326, c.1949
WOOD, 5 TUBES, 2 BANDS
$45

CGE
Canadian General Electric
C352, c.1951
WOOD, 6 TUBES, 1 BAND
$45

CGE
Canadian General Electric
C400, c.1951
BAKELITE, 4 TUBES, 1 BAND
$115

CGE
Canadian General Electric
C401, c.1952
BAKELITE, 4 TUBES, 1 BAND
$90

CGE
Canadian General Electric
C402, c.1950
BAKELITE, 5 TUBES, 1 BAND
$55

CGE
Canadian General Electric
C403, c.1953
PLASTIC, 5 TUBES, 1 BAND
$45

CGE
Canadian General Electric
C404, c.1952
BAKELITE, 5 TUBES, 1 BAND
$55

CGE
Canadian General Electric
C405, c.1954
BAKELITE, 5 TUBES, 1 BAND
$65

CGE
Canadian General Electric
C406, c.1954
BAKELITE, 5 TUBES, 1 BAND
$40

CGE
Canadian General Electric
C407, c.1954
BAKELITE, 4 TUBES, 1 BAND
$25

CGE
Canadian General Electric
C408, c.1954
PLASTIC, 5 TUBES, 1 BAND
$45

CGE
Canadian General Electric
C409, c.1954
BAKELITE, 6 TUBES, 1 BAND
$40

CGE
Canadian General Electric
C409A, c.1954
BAKELITE, 6 TUBES, 1 BAND
$40

CGE
Canadian General Electric
C412, c.1956
PLASTIC, 5 TUBES, 1 BAND
$50

CGE
Canadian General Electric
C414, c.1956
PLASTIC, 5 TUBES, 1 BAND
$50

CGE
Canadian General Electric
C420, c.1956
PLASTIC, 5 TUBES, 1 BAND
$20

CGE
Canadian General Electric
C451, c.1951
BAKELITE, 6 TUBES, 1 BAND
$25

CGE
Canadian General Electric
C452, c.1951
WOOD, 6 TUBES, 1 BAND
$20

CGE
Canadian General Electric
C453, c.1952
PLASTIC, 6 TUBES, 1 BAND
$30

CGE
Canadian General Electric
C504, c.1952
PLASTIC, 4 TUBES, 1 BAND
$25

CGE
Canadian General Electric
C505, c.1952
PLASTIC, 5 TUBES, 1 BAND
$25

CGE
Canadian General Electric
C544A, C545, c.1952
WOOD, 5 TUBES, 1 BAND
$256

CGE
Canadian General Electric
C600, c.1952
PLASTIC, 5 TUBES, 1 BAND
$50

CGE
Canadian General Electric
C601, c.1952
PLASTIC, 5 TUBES, 1 BAND
$25

CGE
Canadian General Electric
C602, c.1952
PLASTIC, 5 TUBES, 1 BAND
$25

CGE
Canadian General Electric
C605, c.1952
PLASTIC, 4 TUBES, 1 BAND
$60

CGE
Canadian General Electric
C625, c.1955
PLASTIC, 4 TUBES, 1 BAND
$175

CGE
Canadian General Electric
C640, c.1955
PLASTIC, 5 TUBES, 1 BAND
$55

CGE
Canadian General Electric
C753, c.1952
WOOD, 5 TUBES, 1 BAND
$25

CGE
Canadian General Electric
C850, c.1950
BAKELITE, 5 TUBES, 1 BAND
$40

CGE
Canadian General Electric
C851, c.1952
WOOD, 5 TUBES, 1 BAND
$30

CGE
Canadian General Electric
CT102, c.1958
PLASTIC, 4 TUBES, 1 BAND
$25

CGE
Canadian General Electric
CT103, c.1959
PLASTIC, 6 TUBES, 1 BAND
$65

CGE
Canadian General Electric
CT104, c.1959
PLASTIC, 5 TUBES, 1 BAND
$30

CGE
Canadian General Electric
CT105, c.1959
PLASTIC, 5 TUBES, 1 BAND
$30

CGE
Canadian General Electric
CT106, c.1959
PLASTIC, 4 TUBES, 1 BAND
$25

CGE
Canadian General Electric
CT140, c.1959
PLASTIC, 5 TUBES, 1 BAND
$25

CGE
Canadian General Electric
CXB353, c.1954
PLASTIC, 4 TUBES, 1 BAND
$40

CGE
Canadian General Electric
F4B, c.1937
WOOD, 4 TUBES, 1 BAND, DC
$65

CGE
Canadian General Electric
F5B, c.1937
WOOD, 5 TUBES, 1 BAND, DC
$75

CGE
Canadian General Electric
F6B, c.1937
WOOD, 6 TUBES, 3 BANDS, DC
$115

CGE
Canadian General Electric
F40, c.1937
WOOD, 4 TUBES, 1 BAND
$90

CGE
Canadian General Electric
F55B, c.1937
WOOD, 5 TUBES, 3 BANDS, DC
$110

CGE
Canadian General Electric
F77, c.1937
WOOD, 7 TUBES, 3 BANDS
$125

CGE
Canadian General Electric
F82, c.1937
WOOD, 8 TUBES, 3 BANDS
$150

CGE
Canadian General Electric
G4B, c.1938
WOOD, 4 TUBES, 1 BAND, DC
$65

CGE
Canadian General Electric
G5B, c.1938
WOOD, 5 TUBES, 3 BANDS, DC
$115

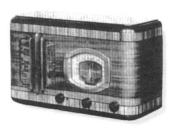

CGE
Canadian General Electric
G6B, c.1938
WOOD, 6 TUBES, 1 BANDS, DC
$90

CGE
Canadian General Electric
G40, c.1938
WOOD, 4 TUBES, 1 BAND
$100

CGE
Canadian General Electric
G51, c.1938
WOOD, 5 TUBES, 2 BANDS
$90

CGE
Canadian General Electric
G60, c.1938
WOOD, 6 TUBES, 1 BAND
$75

CGE
Canadian General Electric
G71, c.1938
WOOD, 7 TUBES, 3 BANDS
$120

CGE
Canadian General Electric
H4B, c.1939
WOOD, 4 TUBES, 1 BAND
$150

CGE
Canadian General Electric
H6B, c.1939
WOOD, 6 TUBES, 1 BAND
$50

CGE
Canadian General Electric
H41B, c.1939
WOOD, 4 TUBES, 1 BAND
$150

CGE
Canadian General Electric
H50, H51, c.1939
WOOD, 5 TUBES, 1 BAND
$75

CGE
Canadian General Electric
H52, c.1939
WOOD, 5 TUBES, 1 BAND
$90

CGE
Canadian General Electric
H53, c.1939
WOOD, 5 TUBES, 2 BANDS
$75

CGE
Canadian General Electric
H61B, H61VB, c.1939
WOOD, 6 TUBES, 3 BANDS
$75

CGE
Canadian General Electric
H70, BANDSPREAD, c.1939
WOOD, 7 TUBES, 5 BANDS
$225

CGE
Canadian General Electric
JK4B, c.1940
WOOD, 4 TUBES, 1 BAND, DC
$50

CGE
Canadian General Electric
JK6B, c.1940
WOOD, 6 TUBES, 1 BAND, DC
$70

CGE
Canadian General Electric
JK40B, 'ACE', c.1940
BAKELITE, 4 TUBES, 1 BAND
$45

CGE
Canadian General Electric
JK44BP, c.1940
METAL, 4 TUBES, 1 BAND, DC
$35

CGE
Canadian General Electric
JK50, c.1940
WOOD, 5 TUBES, 1 BAND
$110

CGE
Canadian General Electric
JK51, c.1940
WOOD, 5 TUBES, 1 BAND
$75

CGE
Canadian General Electric
JK52, c.1940
WOOD, 5 TUBES, 1 BAND
$95

CGE
Canadian General Electric
JK53, c.1940
WOOD & CHROME, 5 TUBES, 3 BANDS
$250

CGE
Canadian General Electric
JK54, c.1940
WOOD, 5 TUBES, 2 BANDS
$85

CGE
Canadian General Electric
JK60B, c.1940
WOOD, 6 TUBES, 5 BANDS
$75

CGE
Canadian General Electric
JK70, c.1940
WOOD, 7 TUBES, 6 BANDS
$110

CGE
Canadian General Electric
K5, RADIO-PHONO, c.1941
WOOD, 5 TUBES, 1 BAND
$50

CGE
Canadian General Electric
KL5, RADIO-PHONO, c.1941
WOOD, 5 TUBES, 4 BANDS
$70

CGE
Canadian General Electric
KL4B, c.1941
WOOD, 4 TUBES, 1 BAND, DC
$30

CGE
Canadian General Electric
KL4B, c.1941
WOOD, 4TUBES, 1 BAND, DC
$35

CGE
Canadian General Electric
KL50, c.1941
WOOD, 5TUBES, 1 BAND
$110

CGE
Canadian General Electric
KL51A, c.1941
WOOD, 5TUBES, 1 BAND
$45

CGE
Canadian General Electric
KL52, c.1941
BAKELITE, 5TUBES, 1 BAND
$95

CGE
Canadian General Electric
KL53, c.1941
WOOD & CHROME, 5 TUBES, 3 BAND
$250

CGE
Canadian General Electric
KL60, c.1941
WOOD, 6 TUBES, 4 BANDS
$90

CGE
Canadian General Electric
KL60B, c.1941
WOOD, 6 TUBES, 5 BANDS, DC
$45

CGE
Canadian General Electric
KL70, c.1941
WOOD, 7 TUBES, 6 BANDS
$175

CGE
Canadian General Electric
KL500, c.1941
WOOD, 5 TUBES, 4 BANDS
$65

CGE
Canadian General Electric
R93AG, PHONO REMOTE, c.1939
BAKELITE
$50

CGE
Canadian General Electric
T145A, c.1960
PLASTIC, TRANSISTORS
$70

CGE
Canadian General Electric
X104CAB, c.1942
WOOD, 4 TUBES, 3 BANDS, DC
$30

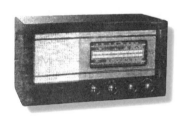

CGE
Canadian General Electric
X115CA, X125CA, c.1942
WOOD, 5 TUBES, 3 BANDS
$40

CHANNEL MASTER
6511, c.1959
PLASTIC, TRANSISTORS
$75

Clarion
404, c.1934
WOOD, 4 TUBES, 1 BAND
$135

Clarion
423, c.1934
WOOD, 5 TUBES, 1 BAND
$145

Clarion
11802, c.1948
BAKELITE, 5 TUBES, 1 BAND
$60

Climax
'CONQUISTADOR', c.1935
WOOD, 5 TUBES, 1 BAND
$275

Climax
256, c.1937
METAL, 5 TUBES, 1 BAND
$125

Climax
'CHICADEE', c.1937
WOOD, 4 TUBES, 1 BAND
$250

Climax
'Emerald'(type b), c.1937
WOOD, 6 TUBES, 2 BANDS
$850

Climax
'Ruby', c.1937
WOOD, 6 TUBES, 2 BANDS
$1,500

Cᴏʟᴏɴɪᴀʟ
136, c.1933
WOOD, 4 TUBES, 1 BAND
$150

Cᴏʟᴏɴɪᴀʟ
400, c.1933
WOOD, 6 TUBES, 1 BAND
$375

Cᴏʟᴏɴɪᴀʟ
651, c.1933
WOOD, 5 TUBES, 2 BANDS
$225

Cᴏʟᴏɴɪᴀʟ
652, c.1933
WOOD, 5 TUBES, 2 BANDS
$225

Cᴏʟᴏɴɪᴀʟ
653, c.1933
WOOD, 5 TUBES, 1 BAND
$375

Columbia
c.1935
BAKELITE
$350

Columbia
2160, c.1958
PLASTIC, 5 TUBES, 1 BAND
$75

Commonwealth
'Compact 6', c.1935
WOOD, 6 TUBES, 1 BAND
$175

Companionette
'Deluxe Chest', c.1933
WOOD
$225

Companionette
'Personal', c.1933
WOOD
$85

Con-Rad

660, c.1935
WOOD, 6 TUBES, 1 BAND
$175

Consolette

'JR.', c.1931
WOOD
$325

Constructorad

1-TUBE RADIO KIT, c.1934

$150

Continental

5S, c.1935
WOOD, 5 TUBES, 2 BANDS
$110

Continental

5SW, c.1935
5 TUBES, 2 BANDS
$150

Continental

14, c.1935
WOOD, 4 TUBES, 1 BAND
$75

CORDONIC

'MIDGET', c.1931
WOOD & TAPESTRY
$500

Coronado

504, c.1938
WOOD, 5 TUBES, 1 BAND, DC
$115

Coronado

521, c.1938
WOOD, 5 TUBES, 1 BAND
$75

Coronado

541A, c.1938
WOOD, 4 TUBES, 1 BAND, DC
$35

Coronado

675A, c.1938
WOOD, 5 TUBES, 1 BAND, DC
$35

Coronado

'JEWEL', c.1938
BAKELITE, 4 TUBES, 1 BAND
$275

COSMAN

'COSMAN TWO', C.1934
METAL, 2 TUBES, SHORTWAVE, DC
$225

CROSLEY

5F, C.1955
BAKELITE, 5 TUBES, 1 BAND
$65

CROSLEY

M3, C.1935
WOOD & CHROME, 5 TUBES, 1 BAND
$125

CROSLEY

5V4, C.1934
WOOD & CHROME, 5 TUBES, 1 BAND
$325

CROSLEY

11-106, C.1053
BAKELITE
$90

CROSLEY

11-108, C.1953
BAKELITE
$70

CROSLEY

52-FC, C.1942
WOOD, 5 TUBES, 1 BAND
$55

CROSLEY

52-TA, C.1942
WOOD, 5 TUBES, 1 BAND
$60

CROSLEY

52-TL, C.1942
WOOD, 5 TUBES, 1 BAND
$100

CROSLEY

56-TH
BAKELITE & ACRYLIC
$65

CROSLEY

615C, 'CRUISER', C.1936
WOOD, 6 TUBES, 3 BANDS
$145

CROSLEY

163, C.1934
WOOD, 5 TUBES, 1 BAND
$225

CROSLEY

181, c.1934

WOOD & CHROME, 6 TUBES, 2 BANDS

$425

CROSLEY

415AA, 'BATTERY 4', c.1936

WOOD & CHROME, 4 TUBES, 1 BAND, DC

$125

CROSLEY

448A, RADIO-PHONO, c.1939

WOOD, 4 TUBES, 1 BAND

$115

CROSLEY

515AC, 'FIVER', c.1936

WOOD, 5 TUBES, 2 BANDS

$100

CROSLEY

517-6K, c.1939

WOOD, 5 TUBES, 1 BAND

$325

CROSLEY

525B, 'GALLEON', c.1936

WOOD, 5 TUBES, 2 BANDS

$175

CROSLEY

555KC, 'BATTERY 5', c.1936

WOOD & CHROME, 5 TUBES, 2 BANDS, DC

$150

CROSLEY

557, c.1940

WOOD, 6 TUBES, 2 BANDS, DC

$75

CROSLEY

625E, 'BATTERY 6', c.1936

WOOD, 6 TUBES, 3 BANDS, DC

$150

CROSLEY

628B, 5628A,B, c.1939

BAKELITE, 6 TUBES, 2 BANDS

$110

CROSLEY

635C, 'BUCCANEER', c.1936

WOOD, 6 TUBES, 3 BANDS

$165

CROSLEY

638B, c.1939

WOOD, 6 TUBES, 2 BANDS

$80

CROSLEY

645CB, c.1936
WOOD, 6 TUBES, 2 BANDS, DC
$90

CROSLEY

648A(BROWN),648B(IVORY),
648C (RED), c.1939
BAKELITE, 6 TUBES, 1 BANDS
$145

CROSLEY

715D, 'CORSAIR', 1936
WOOD, 7 TUBES, 3 BANDS
$195

CROSLEY

725D, 'VIKING', 1936
WOOD, 7 TUBES, 5 BANDS
$250

CROSLEY

758A, HIFI, c.1939
WOOD, 7 TUBES, 2 BANDS
$225

CROSLEY

818A, c.1939
WOOD, 8 TUBES, 3 BANDS
$175

CROSLEY

855F, 'MERRIMAC', c.1936
WOOD, 8 TUBES, 3 BANDS
$325

CROSLEY

865F, 'MONITOR', c.1936
WOOD, 8 TUBES, 5 BANDS
$350

CROSLEY

915EK, 'CLIPPER', c.1936
WOOD, 9 TUBES, 5 BANDS
$500

CROSLEY

1055EK, 'CLIPPER', c.1936
WOOD, 10 TUBES, 5 BANDS
$650

CROSLEY

B46, c.1947
WOOD, 4 TUBES, 1 BAND, DC
$75

CROSLEY

B375, c.1937
WOOD, 4 TUBES, 1 BAND, DC
$95

CROSLEY

B445, c.1937
WOOD, 5 TUBES, 2 BANDS, DC
$100

CROSLEY

B468A, c.1939
WOOD, 4 TUBES, 1 BAND, DC
$50

CROSLEY

B554A, c.1939
WOOD, 5 TUBES, 1 BAND, DC
$75

CROSLEY

B695, c.1937
WOOD, 6 TUBES, 3 BANDS, DC
$100

CROSLEY

C568A,B (BROWN, IVORY), c.1939
BAKELITE, 5 RUBES, 1 BAND
$90

CROSLEY

C648D, c.1939
WOOD, 6 TUBES, 1 BAND
$110

CROSLEY

'DUAL 60', c.1934
WOOD & CHROME, 6 RUBES, 2 BANDS
$425

CROSLEY

E-75, c.1953
PLASTIC
$45

CROSLEY

E-85, c.1953
PLASTIC
$45

CROSLEY

E-90, c.1953
PLASTIC
$45

CROSLEY

F118B, READO PRINTER, c.1939
WOOD, FACSIMILE RECEIVER
$500

CROSLEY

'FORTY', c.1934
WOOD & CHROME, 4 TUBES, 1 BAND
$350

CROSLEY

'NEW FIVER', C.1934

WOOD & CHROME, 4 TUBES, 1 BAND

$225

CROSLEY

'RADIO-TIMER', C.1939

METAL

$200

CROSLEY

'SKYMASTER', C.1953

PLASTIC

$65

CROSLEY

'SKYROCKET', C.1953

PLASTIC

$65

CROSLEY

'SUPER 8', C.1938

WOOD, 8 TUBES, 3 BANDS

$145

CROSLEY

'TRAVETTE MODERNE', C.1934

WOOD & CHROME, 5 TUBES, 1 BAND

$475

CROSLEY

'TRAVO DELUXE', C.1934

WOOD, 4 TUBES, 1 BAND

$275

CROSLEY

'TRAVO', C.1934

METAL, 4 TUBES, 1 BAND

$150

CROYDON

136X, C.1937

WOOD, 5 TUBES, 2 BANDS

$175

DeForest

C.1930

4 TUBES, SHORTWAVE, DC

$750

DeForest-Crosley

8D725, C.1939

WOOD, 7 TUBES, 3 BANDS

$225

DeForest-Crosley

504, C.1934

WOOD, 5 TUBES, 1 BAND

$325

DELCO

1101, c.1936

WOOD, 5 TUBES, 1 BAND

$130

DELCO

1105, c.1936

WOOD, 5 TUBES, 2 BANDS

$120

DELCO

1106, c.1936

WOOD, 5 TUBES, 2 BANDS

$100

DELCO

1107, c.1936

WOOD, 6 TUBES, 3 BANDS

$110

DELCO

1125, c.1938

WOOD, 5 TUBES, 1 BAND

$85

DELCO

1126, c.1938

WOOD, 6 TUBES, 2 BANDS

$75

DELCO

1127, c.1938

WOOD, 6 TUBES, 3 BANDS

$95

DELCO

1128, c.1938

WOOD, 7 TUBES, 3 BANDS

$75

DELCO

1134, c.1938

BAKELITE, 5 TUBES, 1 BAND

$110

DELCO

1174, c.1947

WOOD, 5 TUBES, 1 BAND

$60

DELCO

2055, c.1938

WOOD, 5 TUBES, 1 BAND, DC

$45

DELCO

3201, c.1935

WOOD, 6 TUBES, 1 BAND

$145

DELCO

3203, c.1935

WOOD, 6 TUBES, 2 BANDS

$165

DELCO

3210, c.1938

WOOD, 6 TUBES, 2 BANDS, DC

$40

DELCO

6015, c.1938

WOOD, 6 TUBES, 3 BANDS, DC

$40

DE LUXE

250, c.1934

WOOD, 5 TUBES, 1 BAND

$125

DE LUXE

220, c.1934

WOOD, 5 TUBES, 1 BAND

$125

DE LUXE

'JUNIOR', c.1931

WOOD, 6 TUBES, 1 BAND

$250

DE LUXE

'MIDGET SUPERHET', c.1931

WOOD, 8 TUBES, 1 BAND

$225

DE LUXE

'MULTI-MIDGET', c.1931

WOOD, 6 TUBES, 1 BAND

$225

DETROLA

c.1935

WOOD, 6 TUBES, 2 BANDS

$500

DETROLA

c.1934

WOOD, 4 TUBES, 1 BAND

$300

DETROLA

c.1935

WOOD, 6 TUBES, 2 BANDS

$200

DETROLA

c.1937

WOOD, 6 TUBES, 2 BANDS

$275

DETROLA

6W, c.1936

WOOD, 6 TUBES, 2 BANDS

$275

DETROLA

7A3, c.1935

WOOD, 7 TUBES, 2 BANDS

$225

DETROLA

114, c.1936

WOOD, 6 TUBES, 2 BANDS

$250

DETROLA

140, c.1936

WOOD, 6 TUBES, 2 BANDS

$250

DETROLA

145B, c.1938

WOOD, 7 TUBES, 3 BANDS, DC

$225

DETROLA

162, c.1938

WOOD, 6 TUBES, 2 BANDS

$165

DETROLA

172A, c.1937

WOOD, 5 TUBES, 2 BANDS

$225

DETROLA

180EA, c.1938

WOOD, 5 TUBES, 1 BANDS, DC

$125

DETROLA

199B (BROWN), 199C (IVORY),
c.1939

BAKELITE/PLASKON, 5 TUBES, 1 BAND

BROWN $275

IVORY $450

DETROLA

208AP, RADIO-PHONE,
c.1939

WOOD, 5 TUBES, 1 BAND

$75

DETROLA

209, c.1939

WOOD, 8 TUBES, 3 BANDS

$250

DETROLA

220, c.1939

WOOD, 6 TUBES, 2 BANDS

$225

DETROLA

221, c.1939

WOOD, 6 TUBES, 2 BANDS

$175

DETROLA

222, c.1939

WOOD, 7 TUBES, 2 BANDS

$275

DETROLA

223, c.1939

WOOD, 7 TUBES, 3 BANDS

$275

DETROLA

225, c.1939

WOOD, 7 TUBES, 3 BANDS

$250

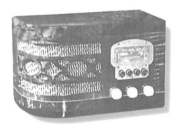

DETROLA

226, c.1939

WOOD, 5 TUBES, 2 BANDS

$225

DETROLA

231, c.1939

WOOD, 9 TUBES, 3 BANDS

$300

DETROLA

248, c.1939

WOOD, 6 TUBES, 2 BANDS

$125

DETROLA

249, c.1939

WOOD, 6 TUBES, 2 BANDS

$150

DETROLA

250, c.1939

WOOD, 7 TUBES, 2 BANDS

$135

DETROLA

251, c.1939

WOOD, 7 TUBES, 3 BANDS

$225

DETROLA

272, "JR.", c.1939

BAKELITE/PLASKON, 4 TUBES, 1 BAND

BROWN $200, BLACK $225

IVORY $325, RED $500

DETROLA

1137, c.1937

WOOD, 7 TUBES, 2 BANDS

$150

DETROLA

1595, c.1938

WOOD, 5 TUBES, 2 BANDS

$125

DETROLA

1795, c.1938

WOOD, 6 TUBES, 2 BANDS

$225

DETROLA

2595, c.1938

WOOD, 8 TUBES, 3 BANDS

$300

DETROLA

3234, c.1938

WOOD, 7 TUBES, 3 BANDS, DC

$150

DETROLA

CM227A, 'BULLET', c.1939

WOOD & BEETLE, 6 TUBES, 2 BANDS

$350

DETROLA

CM254A, 'BULLET', c.1939

WOOD, 6 TUBES, 2 BANDS

$150

DETROLA

'FARM LAYDOWN', c.1938

WOOD, 7 TUBES, 3 BANDS, DC

$125

DETROLA

'FARM UPRIGHT', c.1938

WOOD, 7 TUBES, 3 BANDS, DC

$150

DETROLA

'KENMORE', c.1933

WOOD

$250

DETROLA

'DUAL BAND', c.1933

WOOD, 8 TUBES, 2 BANDS

$600

DETROLA

T2, MOTOR TUNING, c.1938

WOOD, 9 TUBES, 2 BANDS

$650

DETROLA

T3, MOTOR TUNING, c.1938

WOOD, 9 TUBES, 2 BANDS

$650

Diamond

1041XD, c.1935
WOOD, 5 TUBES, 2 BANDS
$275

Diamond

1041XG, c.1935
WOOD, 5 TUBES, 2 BANDS
$225

Diamond

1042PD, c.1935
WOOD, 5 TUBES, 2 BANDS
$250

Diamond

1042PG, c.1935
WOOD, 5 TUBES, 2 BANDS
$300

Diamond

1042U, c.1935
WOOD, 5 TUBES, 2 BANDS
$225

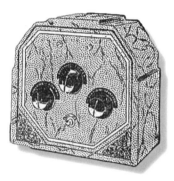

Dixon

SW ADAPTER, c.1932

$125

DUBELIER

'JUNIOR', c.1931
WOOD, 5 TUBES, 1 BAND
$175

DUBELIER

'MIDGET', c.1931
WOOD, 4 TUBES, 1 BAND
$225

DUBELIER

'SENIOR', c.1931
WOOD, 5 TUBES, 1 BAND
$225

Dunlop

c.1939
BAKELITE, 5 TUBES, 1 BAND
$150

Dwarf

1-TUBE KIT, c.1940
WOOD, 1 TUBES, 1 BAND
$125

EDISON-BELL

64, c.1935
WOOD, 6 TUBES, 2 BANDS
$450

EDISON-BELL

65, c.1935
WOOD, 6 TUBES, 2 BANDS
$450

Electrohome

5T7, c.1956
PLASTIC, 5 TUBES, 1 BAND
$25

Electrohome

5T9, c.1956
PLASTIC, 5 TUBES, 1 BAND
$15

Electrohome

51-30, c.1950
PLASTIC, 5 TUBES, 1 BAND
$45

Electrohome

52-05, c.1953
PLASTIC, 5 TUBES, 1 BAND
$20

Electrohome

52-13A, c.1952
PLASTIC, 4 TUBES, 1 BAND
$20

Electrohome

52-40R, c.1952
WOOD, 5 TUBES, 1 BAND
$25

Electrohome

54-13AA, c.1954
PLASTIC, 5 TUBES, 1 BAND
$20

Electrohome

54-17R, c.1954
PLASTIC, 5 TUBES, 1 BAND
$45

Electrohome

54-19, c.1954
PLASTIC, 5 TUBES, 1 BAND
$30

Electrohome

54-20, c.1954
PLASTIC, 5 TUBES, 1 BAND
$25

Electrohome

54-55, c.1954
WOOD, 5 TUBES, 1 BAND
$20

Electrohome
P1O3, c.1953
PLASTIC, 5 TUBES, 1 BAND
$25

Electrohome
P1O7, c.1953
PLASTIC, 5 TUBES, 1 BAND
$35

Electrohome
PK1O4, c.1953
PLASTIC, 5 TUBES, 1 BAND
$35

Electrohome
PMB62-417, c.1950
PLASTIC, 6 TUBES, 1 BAND
$35

Electrohome
PMB62-427, c.1950
PLASTIC, 6 TUBES, 1 BAND
$35

Electrohome
RM274, c.1956
PLASTIC, 6 TUBES, 1 BAND
$25

Electrohome
RM275, c.1956
PLASTIC, 5 TUBES, 1 BAND
$35

Electrohome
RM276, c.1956
PLASTIC, 5 TUBES, 1 BAND
$25

Electrohome
RM278, c.1956
WOOD, 7 TUBES, 3 BANDS
$35

Electrohome
RM221AR, c.1956
PLASTIC, 6 TUBES, 1 BAND
$25

Electrohome
RM237,RM238, c.1956
PLASTIC, 5 TUBES, 1 BAND
$35

Elgin
'MIDGET 5,6 OR 7', c.1932
WOOD, 5,6,7 TUBES, 1 BAND
$250

Emerson

33AW, c.1934

INGRAHAM

WOOD & METAL, 5 TUBES, 2 BANDS

$375

Emerson

117, c.1937

INGRAHAM, 6 TUBES, 1 BANDS, DC

$75

Emerson

126, c.1937

BAKELITE, 5 TUBES, 1 BAND

$150

Emerson

130, c.1937

INGRAHAM, 4 TUBES, 1 BAND, DC

$50

Emerson

131, c.1938

INGRAHAM, 6 TUBES, 1 BAND

$80

Emerson

137, c.1937

INGRAHAM

WOOD & BRASS, 5 TUBES, 1 BAND

$225

Emerson

143, RADIO-PHONO, c.1937

INGRAHAM, 5 TUBES, 1 BAND

$100

Emerson

149, c.1937

BAKELITE/PLASKON, 6 TUBES, 1 BAND

$225

Emerson

156, c.1938

INGRAHAM, 5 TUBES, 1 BAND

$110

Emerson

157, c.1938

BAKELITE/PLASKON, 5 TUBES, 1 BAND

$250

Emerson

158, c.1938

INGRAHAM

WOOD & BRASS, 5 TUBES, 1 BAND

$75

Emerson

166, RADIO-PHONO, c.1938

INGRAHAM, 6 TUBES, 2 BANDS

$275

Emerson

167, C.1938
INGRAHAM, 5 TUBES, 1 BAND
$175

Emerson

168, C.1938
INGRAHAM, 6 TUBES, 1 BAND
$85

Emerson

170, C.1938
INGRAHAM, 6 TUBES, 2 BANDS
$650

Emerson

171, C.1938
INGRAHAM, 6 TUBES, 2 BANDS
$175

Emerson

173, C.1938
INGRAHAM, 6 TUBES, 2 BANDS
$225

Emerson

176, C.1938
INGRAHAM, 6 TUBES, 2 BANDS
$375

Emerson

177, RADIO-PHONO, C.1938
INGRAHAM, 6 TUBES, 2 BANDS
$1,500

Emerson

188, C.1938
BAKELITE/PLASKON, 4 TUBES, 1 BAND
BROWN $300
IVORY $450

Emerson

191, C.1938
BAKELITE/PLASKON, 5 TUBES, 1 BAND
BROWN $75
IVORY $150

Emerson

195, C.1938
INGRAHAM, 6 TUBES, 2 BANDS
$650

Emerson

211, C.1938
BAKELITE/PLASKON, 5 TUBES, 1 BAND
BROWN $100
IVORY $225

Emerson

213, C.1938
INGRAHAM, 5 TUBES, 1 BAND
$950

Emerson

216, RADIO-PHONO, C.1939
INGRAHAM, 5 TUBES, 1 BAND
$50

Emerson

218, RADIO-PHONO, C.1939
INGRAHAM, 5 TUBES, 1 BAND
$150

Emerson

219, RADIO-PHONO, C.1939
INGRAHAM, 5 TUBES, 1 BAND
$50

Emerson

220, RADIO-PHONO, C.1939
INGRAHAM, 5 TUBES, 1 BAND
$70

Emerson

221, RADIO-PHONO, C.1939
INGRAHAM, 5 TUBES, 1 BAND
$150

Emerson

241, RADIO-PHONO, C.1939
INGRAHAM, 5 TUBES, 1 BAND
$250

Emerson

242, RADIO-PHONO, C.1939
INGRAHAM, 5 TUBES, 1 BAND
$50

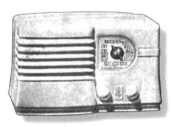

Emerson

246, C.1939
BAKELITE/PLASKON, 5 TUBES, 1 BAND
BROWN $175
IVORY $275
PALE GREEN $600

Emerson

253, C.1939
PRESSED BOARD, 5 TUBES, 1 BAND
$100

Emerson

257, 'KITCHEN', C.1939
INGRAHAM, 5 TUBES, 1 BAND
$250

Emerson

260, C.1939
INGRAHAM, 5 TUBES, 1 BAND, DC
$45

Emerson

312, C.1942
INGRAHAM, 6 TUBES, 1 BAND
$125

Emerson

330, C.1941

BAKELITE, 5 TUBES, 1 BAND

$65

Emerson

332, C.1941

INGRAHAM, 6 TUBES, 2 BANDS

$85

Emerson

333, C.1941

BAKELITE, 5 TUBES, 1 BAND

$65

Emerson

334, C.1941

INGRAHAM, 5 TUBES, 1 BAND

$65

Emerson

337, C.1941

BAKELITE, 6 TUBES, 2 BANDS

$65

Emerson

342, C.1941

INGRAHAM, 5 TUBES, 1 BAND

$50

Emerson

344, C.1941

BAKELITE, 5 TUBES, 1 BAND, DC

$55

Emerson

345, RADIO-PHONO, C.1941

INGRAHAM, 5 TUBES, 1 BAND

$50

Emerson

346, RADIO-PHONO, C.1941

INGRAHAM, 6 TUBES, 1 BAND

$50

Emerson

347, C.1941

INGRAHAM, 5 TUBES, 1 BAND

$225

Emerson

348, C.1941

INGRAHAM, 5 TUBES, 1 BAND

$150

Emerson

349, C.1941

INGRAHAM, 5 TUBES, 1 BAND

$225

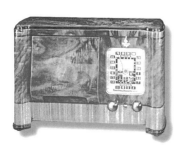

Emerson

352, c.1941
INGRAHAM, 5 TUBES, 1 BAND
$200

Emerson

355, c.1941
INGRAHAM, 5 TUBES, 1 BAND
$75

Emerson

359, c.1941
INGRAHAM, 5 TUBES, 1 BAND, DC
$40

Emerson

364, RADIO-PHONO-RECORDER,
c.1941
INGRAHAM, 7 TUBES, 1 BAND
$175

Emerson

366, c.1941
INGRAHAM, 5 TUBES, 1 BAND
$70

Emerson

367, c.1941
INGRAHAM, 5 TUBES, 1 BAND
$450

Emerson

377, RADIO-PHONO, c.1941
INGRAHAM, 6 TUBES, 1 BAND
$50

Emerson

378, RADIO-PHONO, c.1941
INGRAHAM, 6 TUBES, 1 BAND
$50

Emerson

379, 380, c.1941
PRESSBOARD, 4 TUBES, 1 BAND, DC
$65

Emerson

381, c.1941
BAKELITE/PLASKON, 5 TUBES, 1 BAND
BROWN $75, IVORY $125
RED $500, GREEN $500

Emerson

383, RADIO-PHONO, c.1941
INGRAHAM, 6 TUBES, 1 BAND
$150

Emerson

388, PHONO ONLY, c.1941
INGRAHAM
$30

Emerson

401, c.1942

INGRAHAM, 5 TUBES, 1 BAND

$150

Emerson

406, 'PATRIOT', c.1942

INGRAHAM, 5 TUBES, 1 BAND

$125

Emerson

459, c.1942

INGRAHAM, 6 TUBES, 2 BANDS

$50

Emerson

461, c.1942

PRESSBOARD, 5 TUBES, 1 BAND

$110

Emerson

503, c.1946

WOOD, 5 TUBES, 1 BAND

$60

Emerson

509, c.1946

BAKELITE, 5 TUBES, 1 BAND

$60

Emerson

512, c.1946

WOOD, 6 TUBES, 2 BANDS

$50

Emerson

514, c.1947

BAKELITE, 6 TUBES, 2 BANDS

$50

Emerson

515, c.1946

WOOD, 6 TUBES, 2 BANDS

$60

Emerson

519, c.1946

WOOD, 5 TUBES, 1 BAND

$50

Emerson

524, c.1948

WOOD, 6 TUBES, 4 BANDS

$50

Emerson

528, c.1948

WOOD, 6 TUBES, 2 BANDS

$50

Emerson

530, c.1948

WOOD, 6 TUBES, 2 BANDS

$50

Emerson

531, c.1948

WOOD, 6 TUBES, 2 BANDS, DC

$30

Emerson

532, c.1948

BAKELITE, 6 TUBES, 2 BANDS, DC

$30

Emerson

535, c.1948

WOOD, 5 TUBES, 1 BAND

$50

Emerson

541, c.1948, LOEWY DES.

WOOD, 5 TUBES, 1 BANDS

$50

Emerson

544S, c.1948, BEL GEDDES DES.

WOOD, 6 TUBES, 2 BANDS

$50

Emerson

550, c.1948

WOOD, 6 TUBES, 2 BANDS

$40

Emerson

713, c.1954

WOOD, 5 TUBES, 4 BAND

$90

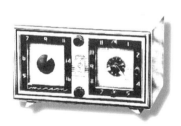

Emerson

718, c.1954

PLASTIC, 5 TUBES, 1 BAND

$65

Emerson

756, c.1954

PLASTIC, 7 TUBES, AMFM

$45

Emerson

779, c.1954

WOOD & PLASTIC, 5 TUBES, 1 BAND

$65

Emerson

790, c.1954

PLASTIC, 5 TUBES, 1 BAND

$85

Emerson

822, C.1955

PLASTIC, 5 TUBES, 1 BAND

$50

Emerson

823, C.1955

PLASTIC, 5 TUBES, 1 BAND

$35

Emerson

822, C.1956

PLASTIC, 5 TUBES, 1 BAND

$110

Emerson

830, C.1956

PLASTIC, 5 TUBES, 1 BAND

$80

Emerson

833, C.1956

PLASTIC, 4 TUBES, 1 BAND

$90

Emerson

1002, C.1948

BAKELITE, 5 TUBES, 1 BAND

$50

Emerson

1700, C.1961

PLASTIC, 5 TUBES, 1 BAND

$35

Emerson

1701, C.1961

PLASTIC, 5 TUBES, 1 BAND

$30

Emerson

1702, C.1961

PLASTIC, 5 TUBES, 1 BAND

$25

Emerson

1703, C.1961

PLASTIC, 5 TUBES, 1 BAND

$75

Emerson

1704, C.1961

PLASTIC, 5 TUBES, 1 BAND

$35

Emerson

1705, C.1961

PLASTIC, 5 TUBES, 1 BAND

$35

Emerson

1706, c.1961
PLASTIC, 5 TUBES, 1 BAND
$35

Emerson

1708, c.1961
PLASTIC, 8 TUBES, AMFM
$75

Emerson

1709, c.1961
PLASTIC, 8 TUBES, AMFM
$45

Emerson

1710, c.1961
PLASTIC, 5 TUBES, 1 BAND
$35

Emerson

1711, c.1961
PLASTIC, 5 TUBES, 1 BAND
$35

Emerson

1713, c.1961
PLASTIC, 5 TUBES, 1 BAND
$35

Emerson

TS, c.1933
WOOD, 5 TUBES, 1 BAND
$125

Emerson

'WONDERGRAM', MINI-PHONO,
c.1959
PLASTIC & BRASS
$225

EMPIRE

450A, c.1935
WOOD, 5 TUBES, 1 BAND
$225

EMPIRE

460B, c.1935
WOOD, 6 TUBES, 2 BANDS
$275

Empress

'BIRDHOUSE', c.1957
WOOD, 5 TUBES, 1 BAND
$450

ESPEY

641, c. 1936
WOOD, 4 TUBES, 1 BAND
$40

ESPEY
671, 'CHINESE LACQUER', C. 1936
WOOD, 7 TUBES, 2 BANDS
$450

ESPEY
771A(WALNUT),771AW(IVORY), C. 1937
WOOD, 7 TUBES, 2 BANDS
$75

ESPEY
771G, C. 1937
LEATHER, 7 TUBES, 2 BANDS
$400

ESPEY
771H, C. 1937
LEATHER & JADE, 7 TUBES, 2 BANDS
$450

ESPEY
961A, C.1936
WOOD, 6 TUBES, 2 BANDS
$75

ESPEY
550U, C.1956
PLASTIC, 5 TUBES, 1 BAND
$55

FADA Radio
C.1938
WOOD
$225

FADA Radio
C.1938
WOOD
$125

FADA Radio
5F50AG, C.1939
CATALIN, 5 TUBES, 1 BAND
$1200+

FADA Radio
5F50T, C.1939
WOOD, 5 TUBES, 1 BAND
$175

FADA Radio
5F50V,W, C.1939
BAKELITE/PLASKON, 5 TUBES, 1 BAND
BROWN $75
IVORY $150

FADA Radio
5F60AG, C.1939
CATALIN, 5 TUBES, 1 BAND
$1200+

FADA *Radio*

5F6OT, c.1939
WOOD, 5 TUBES, 1 BAND
$125

FADA *Radio*

5F6OW, c.1939
BAKELITE, 5 TUBES, 1 BAND
$75

FADA *Radio*

6A6OW, c.1939
BAKELITE, 6 TUBES, 2 BAND
$145

FADA *Radio*

6A65T, c.1939
BAKELITE, 6 TUBES, 2 BAND
$175

FADA *Radio*

12FS, c.1941
WOOD, 5 TUBES, 1 BAND
$40

FADA *Radio*

20V, c.1939
BAKELITE/PLASKON, 6 TUBES, 1 BAND
BROWN $175, BLACK $225,
IVORY $300, RED $1200

FADA *Radio*

30J, c.1939
WOOD, 6 TUBES, 2 BAND
$120

FADA *Radio*

31L, c.1939
WOOD, 5 TUBES, 1 BAND
$100

FADA *Radio*

106, c.1933
PRESSBOARD, 5 TUBES, 1 BAND
$100

FADA *Radio*

107, c.1933
WOOD, 5 TUBES, 1 BAND
$150

FADA *Radio*

182G, 'BABY GRAND',
c.1941
GOLD-PLATED METAL, 5 TUBES, 1 BAND
$750

FADA *Radio*

182S, 'SPINET', c.1941
GOLD-PLATED METAL, 5 TUBES, 1 BAND
$750

FADA *Radio*
29OT, 'STREAMLINE', C.1937
WOOD, 9 TUBES, 4 BANDS
$1,500

FADA *Radio*
451PT, RADIO-PHONO, C.1939
WOOD, 5 TUBES, 1 BAND
$100

FADA *Radio*
461K, C.1939
WOOD, 6 TUBES, 2 BANDS
$110

FADA *Radio*
461L, C.1939
WOOD, 6 TUBES, 2 BANDS
$125

FADA *Radio*
461T, C.1939
WOOD, 6 TUBES, 2 BANDS
$150

FADA *Radio*
465PT, RADIO-PHONO, C.1939
WOOD, 6 TUBES, 2 BANDS
$125

FADA *Radio*
465T, C.1939
WOOD, 6 TUBES, 2 BANDS
$300

FADA *Radio*
465TR, C.1939
WOOD, 6 TUBES, 2 BANDS
$140

FADA *Radio*
465V, C.1939
BAKELITE/PLASKON, 6 TUBES, 2 BANDS
BROWN $225, BLACK $255
IVORY $350, RED $1500

FADA *Radio*
554PT, RADIO-PHONO, C.1939
WOOD, 5 TUBES, 1 BAND
$100

FADA *Radio*
1242PB, C.1939
WOOD, 6 TUBES, 2 BANDS, DC
$110

FADA *Radio*
L56A, C.1939
BAKELITE, 5 TUBES, 1 BAND
$175

FADA *Radio*

RA, c.1934
WOOD, 6 TUBES, 1 BAND
$300

FADA *Radio*

S46B, 'Silent Radio', c.1939
BAKELITE, 5 TUBES, 1 BAND
$225

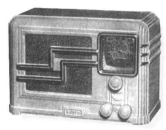

FADA *Radio*

S46G, 'Silent Radio', c.1939
PLASKON & METAL, 6 TUBES, 1 BAND
$450

FADA *Radio*

S46T, 'Silent Radio', c.1939
WOOD, 6 TUBES, 1 BAND
$225

FADA *Radio*

S46W,V,B,R, 'Silent Radio',
c.1939
BAKELITE/PLASKON, 6 TUBES, 1 BAND
BROWN $225, IVORY $350
BLACK $250, RED $1500

FAIRBANKS MORSE

4ATB, c.1938
WOOD, 4 TUBES, 1 BAND, DC
$125

FAIRBANKS MORSE

5AT1, c.1938
WOOD, 5 TUBES, 1 BAND
$150

FAIRBANKS MORSE

5BT2, c.1938
WOOD, 5 TUBES, 2 BANDS
$180

FAIRBANKS MORSE

5CT3, c.1938
WOOD, 5 TUBES, 3 BANDS, DC
$250

FAIRBANKS MORSE

6AT4, c.1938
WOOD, 6 TUBES, 3 BANDS
$250

FAIRBANKS MORSE

6BT6, c.1938
WOOD, 6 TUBES, 2 BANDS
$90

FAIRBANKS MORSE

6CT4B, c.1938
WOOD, 6 TUBES, 3 BANDS, DC
$150

FAIRBANKS MORSE

7AT, C.1939
WOOD, 7 TUBES, 3 BANDS
$275

FAIRBANKS MORSE

8AT8, C.1939
WOOD, 8 TUBES, 3 BANDS
$350

FAIRBANKS MORSE

814, 'SKYSCRAPER', C.1935
WOOD, 8 TUBES, 2 BANDS
$650

FAIRBANKS MORSE

6106, C.1935
WOOD, 5 TUBES, 1 BAND
$225

FAIRBANKS MORSE

5312, 'SKYSCRAPER', C.1935
WOOD, 5 TUBES, 2 BANDS
$350

Farnsworth

BK69, RADIO-PHONO, C.1941
WOOD, 6 RUBES, 2 BANDS
$50

Farnsworth

BT41, C.1941
WOOD, 4 RUBES, 1 BAND, DC
$30

Farnsworth

BT50, C.1941
BAKELITE, 5 TUBES, 1 BAND
$65

Farnsworth

BT52,3, C.1941
BAKELITE, 5 TUBES, 1 BAND
$75

Farnsworth

BT54, C.1941
WOOD, 5 TUBES, 1 BAND
$125

Farnsworth

BT56, C.1941
WOOD, 5 TUBES, 1 BAND
$125

Farnsworth

BT57, C.1941
WOOD, 5 TUBES, 2 BAND, DC
$35

Farnsworth

BT63, c.1941
WOOD, 6 TUBES, 2 BANDS
$85

Farnsworth

BT600, c.1941
WOOD, 6 TUBES, 3 BANDS
$115

Farnsworth

BT1010, c.1941
WOOD, 10 TUBES, 3 BANDS
$175

Farnsworth

CK58, RADIO-PHONO, c.1942
WOOD, 5 TUBES, 1 BAND
$50

Farnsworth

CK66, RADIO-PHONO, c.1942
WOOD, 6 TUBES, 1 BAND
$50

FEDERAL

c.1935
WOOD, 5 TUBES, 1 BAND
$125

FEDERAL

1024TB, c.1948
BAKELITE, 5 TUBES, 1 BAND
$45

FEDERAL

SELECTIVE CRYSTAL, c.1934
WOOD & BAKELITE, CRYSTAL
$250

FEDERAL

3SW, c.1934
METAL, 3 TUBES, SHORTWAVE
$125

FEDERAL

3SW, c.1933
METAL, 3 TUBES, SHORTWAVE
$125

Ferguson

WOOD, 6 TUBES, 2 BANDS, DC
$65

Ferguson

206, c.1938
WOOD, 6 TUBES, 2 BANDS, DC
$85

Firestone

4-A-121, c.1954

BAKELITE, 5 TUBES, 1 BAND

$45

Firestone

4-A-11, c.1950

BAKELITE, 5 TUBES, 1 BAND

$75

Firestone

4-A-12, c.1949

BAKELITE, 5 TUBES, 1 BAND

$150

Firestone

4-A-78, 'EMPIRE', c.1950

BAKELITE, 5 TUBES, 1 BAND

$75

Firestone

4-A-101, c.1954

BAKELITE, 5 TUBES, 1 BAND

$45

Firestone

4-A-108, c.1954

BAKELITE, 5 TUBES, 1 BAND

$65

Firestone

4-A-108, c.1954

WOOD, 5 TUBES, 1 BAND

$50

Firestone

4-A-115, c.1954

WOOD, 5 TUBES, 1 BAND

$55

Firestone

4-A-116, RADIO-PHONO, c.1954

WOOD, 5 TUBES, 1 BAND

$35

Firestone

4-C-17, c.1954

MARBLED PLASTIC, 4 TUBES, 1 BAND

$150

Firestone

4-C-23, c.1954

CANVAS, 4 TUBES, 1 BAND

$25

Firestone

4-C-24, c.1954

CANVAS, 4 TUBES, 1 BAND

$45

Firestone

S-7402-1, c.1939
BEETLE, 5 TUBES, 1 BAND
$275

Firestone

S-7403-3, c.1939
WOOD, 5 TUBES, 1 BAND
$75

Firestone

S-7424-3, c.1939
WOOD, 6 TUBES, 3 BANDS, DC
$50

Firestone

S-7425-6, 'WORLD'S FAIR', c.1939
BAKELITE, 5 TUBES, 1 BAND
$300

Firestone

S-7426-2, c.1939
WOOD, 6 TUBES, 1 BAND
$150

Firestone

S-7426-3, c.1939
WOOD, 6 TUBES, 1 BAND
$150

Firestone

S-7426-5, c.1939
WOOD, 4 TUBES, 1 BAND
$125

Firestone

S-7426-6, c.1939
WOOD, 6 TUBES, 1 BAND
BROWN $75, IVORY $125

Firestone

S-7426-8, c.1939
WOOD, 6 TUBES, 1 BAND
$150

Firestone

S-7427-7, c.1939
WOOD, 5 TUBES, 1 BAND
$75

Firestone

S-7427-8, c.1939
WOOD, 7 TUBES, 2 BANDS
$125

Firestone

S-7428-1, c.1939
BAKELITE/PLASKON
4 TUBES, 1 BAND, DC
$75

Firestone

S-7428-2, c.1939
WOOD, 5 TUBES, 1 BAND, DC
$30

Flewelling

SHORTWAVE ADAPTER, c.1932
WOOD & BAKELITE
$100

Fordson

FY, c.1935
WOOD, 5 TUBES, 1 BAND
$225

Franklin

55GU, c.1935
WOOD, 5 TUBES, 1 BAND
$250

Freed-Eisemann

48, c.1937
WOOD, 4 TUBES, 1 BANDS
$275

Freed-Eisemann

50, c.1937
WOOD, 5 TUBES, 1 BANDS
$275

Freed-Eisemann

70, c.1937
WOOD, 7 TUBES, 2 BANDS
$275

Freshman

MIDGET, c.1933
WOOD, 4 TUBES, 1 BAND
$225

Freshman

MIDGET, c.1933
WOOD, 5 TUBES, 1 BAND
$225

Gamble

MIDGET, c.1933
WOOD, 5 TUBES, 1 BAND
$225

Garod

3P85, RADIO-PHONO, c.1942
WOOD, 8 TUBES, 2 BANDS
$60

Garod

58, c.1935
WOOD, 7 TUBES, 2 BANDS
$225

Garod

66, c.1936

WOOD, 6 TUBES, 1 BAND

$200

Garod

125, c.1941

WOOD, 5 TUBES, 1 BAND

$75

Garod

237, c.1935

WOOD, 7 TUBES, 2 BANDS

$175

Garod

256, c.1941

WOOD, 5 TUBES, 1 BAND

IVORY & IVORY, IVORY & BLUE, IVORY & RED, IVORY
& MAROON, MAROON & IVORY

$900+

Garod

XA-49, c.1949

BAKELITE, 5 TUBES, 1 BAND

$125

GEM

SW ADAPTER, c.1932

1 TUBE, SW

$125

GEM

SW ADAPTER, c.1932

3 TUBES, SW

$175

GENERAL

c.1935

WOOD, 5 TUBES, 1 BAND

$90

GENERAL

521, c.1937

WOOD, 5 TUBES, 1 BAND

$125

GENERAL

622, c.1937

WOOD, 6 TUBES, 1 BAND

$85

GENERAL

1050, c.1934

WOOD & ALUMINUM, 5 TUBES, 1 BAND

$300

GENERAL

MIDGET, c.1932

WOOD, 4 TUBES, 1 BAND

$225

GENERAL ⊛ ELECTRIC

20, C.1942

INGRAHAM, 7 TUBES, 2 BANDS

$175

GENERAL ⊛ ELECTRIC

180, C.1947

WOOD, 4 TUBES, 1 BAND

$50

GENERAL ⊛ ELECTRIC

203, C.1947

WOOD, 5 TUBES, 1 BAND

$50

GENERAL ⊛ ELECTRIC

220, C.1947

BAKELITE, 5 TUBES, 1 BAND

$85

GENERAL ⊛ ELECTRIC

221, C.1947

WOOD, 5 TUBES, 1 BAND

$40

GENERAL ⊛ ELECTRIC

280, C.1947

WOOD, 5 TUBES, 1 BAND

$50

GENERAL ⊛ ELECTRIC

321, C.1947

WOOD, 5 TUBES, 1 BAND

$40

GENERAL ⊛ ELECTRIC

409, C.1951

BAKELITE, 7 TUBES, AMFM

$75

GENERAL ⊛ ELECTRIC

415, C.1952

SWIRLED PLASTIC, 5 TUBES, 1 BAND

$60

GENERAL ⊛ ELECTRIC

422, C.1952

PLASTIC, 6 TUBES, 1 BAND

$40

GENERAL ⊛ ELECTRIC

430, C.1952

PLASTIC, 5 TUBES, 1 BAND

$30

GENERAL ⊛ ELECTRIC

606, C.1951

PLASTIC, 4 TUBES, 1 BAND

$60

GENERAL ELECTRIC

621, C.1955
PLASTIC, 6 TUBES, 1 BAND
$45

GENERAL ELECTRIC

638(GREEN), 639(IVORY), 640(RED),
C.1955
PLASTIC, 6 TUBES, 1 BAND
GREEN $110, IVORY $45, RED $75

GENERAL ELECTRIC

672(BROWN), 673(IVORY), 674(RED),
C.1955
PLASTIC, 6 TUBES, 1 BAND
BROWN $40, IVORY $45, RED $65

GENERAL ELECTRIC

920, C.1955
PLASTIC, 5 TUBES, 1 BAND
$45

GENERAL ELECTRIC

A53, C.1935
WOOD, 5 TUBES, 2 BANDS
$145

GENERAL ELECTRIC

A70, C.1935
WOOD, 7 TUBES, 3 BANDS
$225

GENERAL ELECTRIC

A83, C.1935
WOOD, 8 TUBES, 4 BANDS
$250

GENERAL ELECTRIC

E50, C.1936
WOOD, 5 TUBES, 2 BANDS
WALNUT, IVORY&GOLD, BLACK&GOLD, RED&GOLD
$145

GENERAL ELECTRIC

F40, C.1938
BAKELITE, 4 TUBES, 1 BAND
$125

GENERAL ELECTRIC

F51, C.1938
BAKELITE, 5 TUBES, 2 BANDS
$150

GENERAL ELECTRIC

F53, C.1938
WOOD, 5 TUBES, 2 BANDS
$125

GENERAL ELECTRIC

F74, C.1938
WOOD, 7 TUBES, 2 BANDS
$100

GENERAL ELECTRIC

F80, C.1938

WOOD, 8 TUBES, 3 BANDS

$150

GENERAL ELECTRIC

FD62, C.1938

WOOD, 6 TUBES, 2 BANDS, DC

$75

GENERAL ELECTRIC

H64L, H640L, C.1939

WOOD, 5 TUBES, 3 BANDS

$85

GENERAL ELECTRIC

H406U, 'BANTAM' C.1939

BAKELITE & PLASKON, 4 TUBES, 1 BAND

$225

GENERAL ELECTRIC

H508, RADIO-PHONO, C.1939

WOOD, 5 TUBES, 1 BAND

$65

GENERAL ELECTRIC

H531, C.1939

FAUX LEATHER OVER WOOD

5 TUBES, 1 BAND

$125

GENERAL ELECTRIC

H600, H601, C.1939

BAKELITE, 6 TUBES, 1 BAND

$65

GENERAL ELECTRIC

H610, H611, C.1939

BAKELITE, 6 TUBES, 1 BAND

$75

GENERAL ELECTRIC

H620, H621, C.1939

BAKELITE, 6 TUBES, 1 BAND

$65

GENERAL ELECTRIC

H622, H623, C.1939

WOOD, 6 TUBES, 2 BANDS

$125

GENERAL ELECTRIC

H630, H631, C.1939

BAKELITE, 6 TUBES, 1 BAND

$65

GENERAL ELECTRIC

H632, C.1939

BAKELITE, 6 TUBES, 1 BAND

$75

GENERAL ⊛ ELECTRIC

H638, RADIO-PHONO, C.1939
WOOD, 6 TUBES, 2 BANDS
$90

GENERAL ⊛ ELECTRIC

H639, RADIO-PHONO, C.1939
WOOD, 6 TUBES, 1 BANDS
$125

GENERAL ⊛ ELECTRIC

H640, C.1939
WOOD, 6 TUBES, 1 BAND
$115

GENERAL ⊛ ELECTRIC

HE50, HE540, C.1939
WOOD, 5 TUBES, 3 BANDS
$85

GENERAL ⊛ ELECTRIC

HE74, HE74L, HE740, HE740L,
C.1939
WOOD, 7 TUBES, 3 BANDS
$110

GENERAL ⊛ ELECTRIC

HE100, GHE100L, HE100H,
HE100LH, C.1939
WOOD, 10 TUBES, 3 BANDS
$175

GENERAL ⊛ ELECTRIC

HJ618, RADIO-PHONO, C.1940
WOOD, 6 TUBES, 1 BANDS
$125

GENERAL ⊛ ELECTRIC

HJ619, RADIO-PHONO, C.1940
WOOD/IVORY, 6 TUBES, 1 BANDS
$125

GENERAL ⊛ ELECTRIC

HM3, HM21, WIRELESS PHONO,
C.1939
BAKELITE(VASSOS DES.)
$150

GENERAL ⊛ ELECTRIC

HP558, C.1939
WOOD, 5 TUBES, 1 BAND
$60

GENERAL ⊛ ELECTRIC

HP559, C.1939
WOOD, 5 TUBES, 1 BAND
$65

GENERAL ⊛ ELECTRIC

HP657A, C.1939
WOOD, 6 TUBES, 1 BAND
$65

GENERAL ⊕ ELECTRIC

J44(BROWN), J644W(IVORY), C.1941
BAKELITE/PLASKON, 6 TUBES, 1 BAND
BROWN $65, IVORY $110

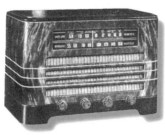

GENERAL ⊕ ELECTRIC

J63, C.1941
INGRAHAM, 6 TUBES, 2 BANDS
$80

GENERAL ⊕ ELECTRIC

J71, C.1940
INGRAHAM, 7 TUBES, 3 BANDS
$120

GENERAL ⊕ ELECTRIC

J602, J603, C.1941
BAKELITE, 6 TUBES, 1 BAND
$50

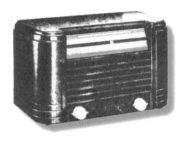

GENERAL ⊕ ELECTRIC

J614, J664, C.1941
BAKELITE, 6 TUBES, 1 BAND
$50

GENERAL ⊕ ELECTRIC

J629, RADIO-PHONO-RECORDER, C.1941
INGRAHAM, 6 TUBES, 1 BAND
$125

GENERAL ⊕ ELECTRIC

J644, J654, C.1941
BAKELITE, 6 TUBES, 1 BAND
$65

GENERAL ⊕ ELECTRIC

JE58, C.1940
WOOD, 5 TUBES, 3 BANDS
$60

GENERAL ⊕ ELECTRIC

JE61, JE61L, C.1940
WOOD, 6 TUBES, 3 BANDS
$60

GENERAL ⊕ ELECTRIC

JE81, JE810, C.1940
WOOD, 8 TUBES, 3 BANDS
$80

GENERAL ⊕ ELECTRIC

JE101, C.1940
WOOD, 10 TUBES, 3 BANDS
$125

GENERAL ⊕ ELECTRIC

JE51, JE510, C.1940
WOOD, 5 TUBES, 3 BANDS
$60

GENERAL ELECTRIC

JE530, JE531, JE531X,
c.1940

WOOD, 5 TUBES, 3 BANDS

$60

GENERAL ELECTRIC

JM3, PHONO REMOTE, c.1940
BAKELITE (VASSOS DES.)

$150

GENERAL ELECTRIC

JM4, JM23, PHONO REMOTE, c.1940
BAKELITE (VASSOS DES.)

$50

GENERAL ELECTRIC

JZ822, c.1932

WOOD, 8 TUBES, 1 BAND

$150

GENERAL ELECTRIC

K50A, c.1933

WOOD, 5 TUBES, 1 BAND

$165

GENERAL ELECTRIC

L50, c.1933

REPWOOD, 5 TUBES, 1 BAND

$375

GENERAL ELECTRIC

L500, L510, L550, L560, c.1941
BAKELITE, 5 TUBES, 1 BAND

$40

GENERAL ELECTRIC

L512, c.1941

BAKELITE, 5 TUBES, 1 BAND

$40

GENERAL ELECTRIC

L540, L541, c.1941

WOOD, 5 TUBES, 1 BAND

$45

GENERAL ELECTRIC

L542, c.1941

INGRAHAM, 5 TUBES, 1 BAND

$65

GENERAL ELECTRIC

L604, c.1942

BAKELITE, 6 TUBES, 1 BAND

$40

GENERAL ELECTRIC

L631, c.1941

WOOD, 6 TUBES, 1 BAND

$45

GENERAL ELECTRIC

L641, c.1941
WOOD, 6 TUBES, 1 BAND
$45

GENERAL ELECTRIC

L642, c.1941
INGRAHAM, 6 TUBES, 1 BAND
$110

GENERAL ELECTRIC

L643, L663, c.1941
WOOD, 6 TUBES, 1 BAND
$45

GENERAL ELECTRIC

L651, c.1941
WOOD, 6 TUBES, 1 BAND
$45

GENERAL ELECTRIC

L652, c.1941
WOOD, 6 TUBES, 1 BAND
$45

GENERAL ELECTRIC

L653, L673, c.1941
WOOD, 6 TUBES, 1 BAND
$40

GENERAL ELECTRIC

L674, c.1940
INGRAHAM, 6 TUBES, 2 BANDS
$90

GENERAL ELECTRIC

L678, RADIO-PHONO, c.1941
WOOD, 6 TUBES, 1 BAND
$95

GENERAL ELECTRIC

L740, c.1941
WOOD, 7 TUBES, 3 BANDS
$85

GENERAL ELECTRIC

LB424, c.1940
WOOD, 4 TUBES, 1 BAND, DC
$35

GENERAL ELECTRIC

LB502, c.1942
TENITE, 5 TUBES, 1 BAND
$110

GENERAL ELECTRIC

LC679, RADIO-PHONO, c.1941
WOOD, 6 TUBES, 1 BAND
$95

GENERAL ELECTRIC

LM20, PHONO REMOTE,
C.1940
INGRAHAM, PHONO ONLY
$95

GENERAL ELECTRIC

LM21, PHONO REMOTE,
C.1940
INGRAHAM, PHONO ONLY
$65

GENERAL ELECTRIC

T16A, C.1956
PLASTIC, 5 TUBES, 1 BAND
$55

GENERAL ELECTRIC

X105A, X105B, X105VB, C.1942
WOOD, 5 TUBES, 3 BANDS
$40

GENERAL ELECTRIC

X108, X118, C.1942
WOOD, 8 TUBES, 7 BANDS
$115

GENERAL ELECTRIC

X115, X125, C.1942
WOOD, 5 TUBES, 3 BANDS
$40

GENERAL ELECTRIC

X145, RADIO-PHONO, C.1942
WOOD, 5 TUBES, 3 BANDS
$65

GENERAL ELECTRIC

X156, X166, C.1942
WOOD, 6 TUBES, 7 BANDS
$55

GENERAL ELECTRIC

X216A, X216V, C.1942
WOOD, 6 TUBES, 7 BANDS
$55

GENERAL ELECTRIC

X225A, X225V, C.1942
WOOD, 65 TUBES, 2 BANDS
$45

GENERAL ELECTRIC

X226, C.1942
WOOD, 6 TUBES, 7 BANDS
$50

GENERAL ELECTRIC

X228, C.1942
WOOD, 8 TUBES, 7 BANDS
$50

GENERAL ☼ ELECTRIC
XFM1, C.1949
WOOD, FM ONLY
$45

General
TELEVISION
522, C.1940
BAKELITE, 5 TUBES, 1 BAND
$225

General
TELEVISION
526, C.1940
WOOD, 5 TUBES, 1 BAND
$150

General
TELEVISION
585, C.1940
WOOD, 5 TUBES, 1 BAND
$150

General
TELEVISION
588, 'TORPEDO', C.1940
BAKELITE, 5 TUBES, 1 BAND
$250

General
TELEVISION
589, C.1940
WOOD, 5 TUBES, 1 BAND
$125

General
TELEVISION
599, C.1940
WOOD, 5 TUBES, 2 BANDS
$95

General
TELEVISION
616, C.1940
WOOD, 6 TUBES, 1 BAND
$60

General
TELEVISION
696, C.1940
BAKELITE, 6 TUBES, 1 BAND
$225

General
TELEVISION
'C', C.1935
WOOD & CHROME, 4 TUBES, 1 BAND
$275

General
TELEVISION
'F', C.1935
WOOD, 5 TUBES, 2 BANDS
$165

General
TELEVISION
'K', C.1935
WOOD, 4 TUBES, 1 BAND
$325

Gilfillan

5F, c.1942

BAKELITE, 5 TUBES, 1 BAND

$70

Gilfillan

5TC, RADIO-PHONO, c.1937

WOOD, 5 TUBES, 1 BAND

$95

Gilfillan

6H, c.1942

WOOD, 6 TUBES, 2 BANDS

$125

Gilfillan

8T, c.1936

WOOD, 8 TUBES, 2 BANDS

$750

Gilfillan

9T, c.1942

WOOD, 9 TUBES, 2 BANDS

$185

Gilfillan

15F(IVORY), 15G(BROWN), c.1942

BAKELITE/PAINTED

5 TUBES, 1 BAND

$120

Gilfillan

15H, c.1942

WOOD, 5 TUBES, 1 BAND

$75

Gilfillan

25H, c.1942

WOOD, 5 TUBES, 1 BAND

$50

Gilfillan

42A, c.1936

WOOD, 4 TUBES, 1 BAND

$115

Gilfillan

43A, c.1936

WOOD, 4 TUBES, 1 BAND

$125

Gilfillan

52A, 53A, c.1936

MIRROR, 5 TUBES, 1 BAND

$1,100+

Gilfillan

54A, 55A, c.1936

WOOD, 5 TUBES, 1 BAND

$450

Gilfillan

62B, 62X, c.1936
WOOD, 6 TUBES, 2 BANDS
$200

Gilfillan

63B, 63X, c.1936
WOOD, 6 TUBES, 3 BANDS
$225

Gilfillan

96B, 96X, c.1936
WOOD, 9 TUBES, 3 BANDS
$350

Gilfillan

116B, c.1936
WOOD, 11 TUBES, 3 BANDS
$450

Gilfillan

402T, c.1937
WOOD, 4 TUBES, 1 BAND
$90

Gilfillan

412T, c.1937
WOOD, 4 TUBES, 1 BAND
$125

Gilfillan

501T, c.1937
WOOD, 5 TUBES, 1 BAND
$175

Gilfillan

504T, c.1937
WOOD, 5 TUBES, 1 BAND
$450

Gilfillan

521T, c.1937
WOOD, 5 TUBES, 2 BANDS
$175

Gilfillan

711T, c.1937
WOOD, 7 TUBES, 2 BANDS
$175

Gilfillan

723T, c.1937
WOOD, 7 TUBES, 2 BANDS, DC
$150

Gilfillan

731T, c.1937
WOOD, 7 TUBES, 3 BANDS
$200

Gilfillan

731T, C.1937
WOOD, 8 TUBES, 3 BANDS
$225

Grunow

510, C.1938
WOOD, 5 TUBES, 2 BANDS, DC
$250

Grunow

550, C.1934
WOOD & CHROME
5 TUBES, 1 BAND
$125

Grunow

566A, C.1938
WOOD, 5 TUBES, 2 BANDS
$135

GREBE

307L, C.1938
WOOD, 7 TUBES, 2 BANDS
$225

hallicrafters

ATX13, C.1954
PLASTIC, 5 TUBES, 1 BAND
$65

hallicrafters

C51, C.1955
WOOD, 5 TUBES, 1 BAND
$110

Halson

Halson

4A, C.1937
WOOD, 6 TUBES, 2 BANDS
$115

Halson

5, C.1934
WOOD, 5 TUBES, SW
$150

Halson

66AW, C.1934
WOOD, 6 TUBES, 2 BANDS
$200

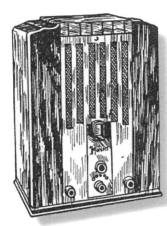

Halson

66AW, C.1934
WOOD, 6 TUBES, 2 BANDS
$250

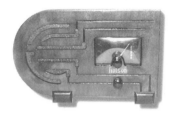

Halson

106, 'BULLET', C.1938
WOOD, 5 TUBES, 1 BANDS
$450

Halson

610, C.1935
WOOD, 6 TUBES, 2 BANDS
$225

Halson

620, C.1935
WOOD, 6 TUBES, 2 BANDS
$275

Halson

630 'DELUXE', C.1935
WOOD, 6 TUBES, 2 BANDS
$275

Halson

630B, C.1935
WOOD, 6 TUBES, 2 BANDS
$175

Halson

770AW, C.1934
WOOD, 5 TUBES, 2 BANDS
$375

Halson

770AW, C.1934
WOOD, 7 TUBES, 2 BANDS
$275

Halson

'MANTLE', C.1933
WOOD, 4 TUBES, 1 BAND
$90

Halson

MIDGET, C.1933
WOOD, 4 TUBES, 1 BAND
$175

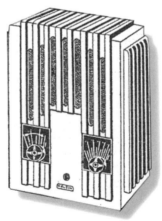

Halson

NS40, C.1934
WOOD, 4 TUBES, 1 BAND
$450

Halson

SUPERHET 6, C.1933
7 TUBES, 1 BAND
$225

Halson

TREASURE CHEST, C.1933
WOOD
$200

HEATHKIT

XR-1L, c.1959
PLASTIC, TRANSISTORS
$90

Hi-lo

TDDA, c.1934
WOOD, 4 TUBES, 1 BAND
$225

Hi-lo

TMMA, c.1934
WOOD, 4 TUBES, 1 BAND
$175

HOWARD

275, c.1938
WOOD
$75

HOWARD

568, c.1940
WOOD, 9 TUBES, 3 BANDS
$75

HOWARD

575, c.1940
WOOD, 6 TUBES, 3 BANDS
$65

HOWARD

580, c.1940
WOOD, 8 TUBES, 3 BANDS
$70

HOWARD

700, c.1941
BAKELITE, 5 TUBES, 1 BAND
$50

HOWARD

B13, c.1935
WOOD, 5 TUBES, 1 BAND
$110

HOWARD

PUSHBUTTON TUNER, c.1937

$50

Hudson

625, c.1936
WOOD, 8 TUBES, 5 BANDS
$225

Hudson

632Y, c.1936
WOOD, 5 TUBES, 2 BANDS
$250

Hudson

634, c.1936
WOOD, 6 TUBES, 3 BANDS
$150

Hudson

CD, c.1936
WOOD, 5 TUBES, 1 BAND
$75

Hudson

MAG, c.1936
WOOD, 4 TUBES, 1 BAND
$60

ICA

'ENVOY', c.1931
WOOD, 5 TUBES, 1 BAND
$250

ICA

'INSULETTE', c.1932
WOOD, 4 TUBES, 1 BAND
$175

ICA

'MARVEL', SW CONVERTER, c.1934
BAKELITE, 1 TUBE
$110

ICA

'MIGHTY MULTI-MU', c.1931
WOOD, 4 TUBES, 1 BAND
$225

ICA

'SCOUT', SW CONVERTER, c.1935
WOOD, 2 TUBES, SW
$75

ICA

'SUPER SW CONVERTER', c.1934
METAL, 1 TUBE, SW
$75

ICA

'TOP NOTCH', c.1940
CRYSTAL
$175

ICA

'TRIODYNE', c.1935
WOOD, 3 BAND
$225

ICA

TWO-TUBE RECEIVER, c.1931
2 TUBES, 1 BAND
$150

Imperial

41A, c.1933
WOOD, 4 TUBES, 1 BAND
$200

Imperial

50, c.1936
WOOD, 5 TUBES, 1 BAND
$115

Imperial

51C, c.1933
WOOD, 5 TUBES, 1 BAND
$200

Imperial

55A, c.1941
WOOD, 7 TUBES, 2 BAND
$95

Imperial

60, c.1936
WOOD, 6 TUBES, 1 BAND
$225

Imperial

71A, c.1933
WOOD, 7 TUBES, 1 BAND
$250

Imperial

75T, c.1934
WOOD, 7 TUBES, 1 BAND
$245

Imperial

104N, c.1941
BAKELITE, 4 TUBES, 1 BAND
$60

Imperial

242, c.1941
WOOD, 5 TUBES, 1 BAND
$95

Imperial

250, c.1934
WOOD, 5 TUBES, 1 BAND
$125

Imperial

4401, c.1934
WOOD, 4 TUBES, 1 BAND
$135

Imperial

540, c.1934
WOOD, 4 TUBES, 1 BAND
$195

Imperial

550, c.1934

WOOD, 5 TUBES, 1 BAND

$185

Imperial

576-5Q, c.1941

BAKELITE, 5 TUBES, 1 BAND

$75

Imperial

626, c.1937

WOOD, 6 TUBES, 2 BANDS

$80

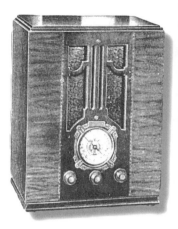

Imperial

626A, c.1936

WOOD, 6 TUBES, 3 BANDS

$160

Imperial

650, c.1934

WOOD, 6 TUBES, 1 BAND

$275

Imperial

694, c.1936

WOOD, 6 TUBES, 3 BANDS

$135

Imperial

1425C, c.1934

METAL, 4 TUBES, 1 BAND

$110

Imperial

1501N, c.1941

BAKELITE, 5 TUBES, 1 BAND

$150

Imperial

1550G, c.1934

WOOD, 5 TUBES, 1 BAND

$350

Imperial

c.1932

WOOD, 7 TUBES, 1 BAND

$425

INSULETTE

MIDGET 5, c.1933

WOOD, 5 TUBES, 1 BAND

$210

INSULETTE

PIED PIER, c.1934

BAKELITE, CRYSTAL

$250

Jackson Bell

86, c.1932

WOOD, 1 BAND

$350

Jackson Bell

206, c.1936

WOOD, 4 TUBES, 1 BAND

$110

Jackson Bell

226, c.1936

WOOD, 4 TUBES, 1 BAND

$110

Jackson Bell

406, c.1936

WOOD, 5 TUBES, 2 BANDS

$90

Jackson Bell

456D, c.1936

WOOD, 5 TUBES, 1 BAND

$135

Jackson Bell

456N, c.1936

WOOD, 5 TUBES, 1 BAND

$110

Jackson Bell

556, c.1936

WOOD, 5 TUBES, 2 BANDS

$175

Jewel

c.1934

WOOD, 4 TUBES, 1 BAND

$95

Jewel

89, 'WAKEMASTER', c.1951

BAKELITE, 5 TUBES, 1 BAND

$60

Jewel

98, 'WAKEMASTER', c.1951

BAKELITE, 5 TUBES, 1 BAND

$70

Jewel

181, c.1951

BAKELITE, 5 TUBES, 1 BAND

$60

Jewel

181A, 'ALWAYS', c.1951

PLASTIC, 4 TUBES, 1 BAND

$60

Jewel

402(BLACK & IVORY), 403(IVORY),
'WAKEMASTER', C.1951
BAKELITE, 5 TUBES, 1 BAND
$60

Jewel

700, C.1951
PLASTIC, 5 TUBES, 1 BAND
$60

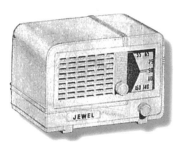

Jewel

400(BROWN), 401(IVORY), C.1951
BAKELITE, 4 TUBES, 1 BAND
BROWN $45, IVORY $70

Jewel

5100S, C.1954
BAKELITE, 4 TUBES, 1 BAND
$45

Jewel

L, C.1935
WOOD, 5 TUBES, 2 BANDS
$375

Jewel

MIDGET, C.1931
WOOD
$225

KADETTE

32, 'FUTURA', C.1936
WOOD, 4 TUBES, 1 BAND
$425

KADETTE

36, C.1936
WOOD, 6 TUBES, 2 BANDS
$65

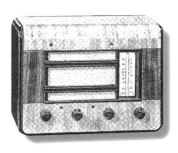

KADETTE

630, C.1938
WOOD, 6 TUBES, 2 BANDS
$65

KADETTE

635, C.1938
WOOD, 6 TUBES, 2 BANDS
$125

KADETTE

735, C.1938
WOOD, 7 TUBES, 2 BANDS
$85

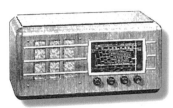

KADETTE

845, C.1938
WOOD, 8 TUBES, 3 BANDS
$110

KADETTE

950, c.1938
WOOD
$115

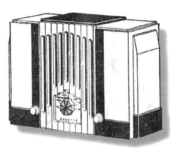

KADETTE

B, c.1933
WOOD, 5 TUBES, 1 BAND
$425

KADETTE

'CONVENTIONAL', c.1933
WOOD, 5 TUBES, 1 BAND
$250

KADETTE

ES, c.1935
WOOD, 5 TUBES, SW ONLY
$300

KADETTE

K10, 'CLASSIC', c.1938
PLASTICS, 6 TUBES, 1 BAND
VARIOUS COLORS $750+

KADETTE

'CLOCKETTE', c.1938
CATALIN, 5 TUBES, 1 BAND
VARIOUS COLORS $1,200+

KADETTE

K40, 'JEWEL', c.1938
BAKELITE & PLASKON
4 TUBES, 1 BAND
$250

KADETTE

'TUNEMASTER', c.1938
WOOD, 3 TUBES
REMOTE CONTROL
$325

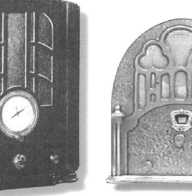

KENNEDY
The Royalty of Radio

5, 5C, c.1935
WOOD, 8 TUBES, 4 BANDS
$300

KENNEDY
The Royalty of Radio

52A, c.1931
WOOD, 6 TUBES, 1 BAND
$350

KENNEDY
The Royalty of Radio

67, c.1933
WOOD, 7 TUBES, 1 BAND
$300

KENNEDY
The Royalty of Radio

610, c.1935
WOOD, 6 TUBES, 3 BANDS
$300

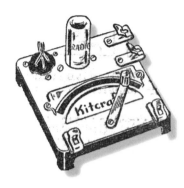

Kitcraft

5001

CRYSTAL

$50

KNIGHT

5F-565, C.1951

PLASTIC

$45

KNIGHT

97-908, C.1951

BAKELITE, POLICE BAND

$35

KNIGHT

101, C. 1942

BAKELITE, 5 TUBE, 1 BAND

$75

KNIGHT

106, C.1942

WOOD, 6 TUBES, 2 BANDS

$45

KNIGHT

108, C.1942

BAKELITE, 6 TUBES, 2 BANDS

$95

KNIGHT

110, C.1942

WOOD, 7 TUBES, 2 BANDS

$55

KNIGHT

111, C.1942

WOOD, 8 TUBES, 3 BANDS

$55

KNIGHT

125, C.1942

WOOD, 7 TUBES, 2 BANDS

$60

KNIGHT

136, FM ADAPTER, C.1942

WOOD, 8 TUBES, FM ONLY

$45

KNIGHT

175, RADIO-PHONO, C.1942

WOOD, 6 TUBE, 3 BAND

$60

KNIGHT

312, C.1942

WOOD, 9 TUBES, 3 BANDS

$60

KNIGHT

325, RADIO-PHONO-RECORDER, C.1942
WOOD, 6 TUBES, 1 BAND
$75

KNIGHT

330, RADIO-PHONO, C.1942
WOOD, 5 TUBES, 1 BAND
$50

KNIGHT

342, C.1942
WOOD, 4 TUBES, 1 BAND, DC
$25

KNIGHT

9516, C.1932
WOOD, 5 TUBES, 1 BAND
$250

KNIGHT

9550, C.1932
WOOD, 6 TUBES, 1 BAND
$350

KNIGHT

9552, C.1932
WOOD, 6 TUBES, 1 BAND
$225

KNIGHT

9947, C.1932
WOOD, 5 TUBES, 1 BAND
$170

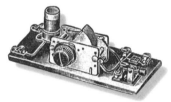

KNIGHT

CRYSTAL KIT, C.1932
CRYSTAL
$60

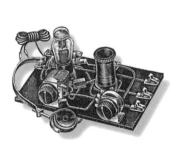

KNIGHT

DX'ER KIT, C.1942
1 TUBE, 1 BAND
$75

KNIGHT

DX'ER KIT, C.1942
2 TUBE, 1 BAND
$75

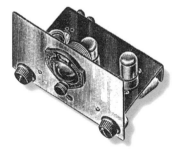

KNIGHT

OCEAN HOPPER, C.1942
2 TUBES, ALL-WAVE
$75

KNIGHT

Y737, RANGER KIT, C.1959
PLASTIC, 5 TUBES, 1 BAND
$45

Lafayette

1-427, c.1953

PLASTIC, 4 TUBE, 1 BAND

$70

Lafayette

1-442, c.1953

PLASTIC, 4 TUBE, 1 BAND

$50

Lafayette

1-443, c.1953

PLASTIC, 4 TUBE, 1 BAND

$40

Lafayette

1-573, c.1953

PLASTIC, 5 TUBES, 1 BAND

$35

Lafayette

1-574, c.1953

WOOD, 5 TUBE, 1 BAND

$50

Lafayette

1N-427, c.1950

PLASTIC, 4 TUBE, 1 BAND

$70

Lafayette

1N-435, c.1950

PLASTIC, 4 TUBE, 1 BAND

$60

Lafayette

1N-551, 'RANGER', c.1950

PLASTIC, 5 TUBE, 1 BAND

$60

Lafayette

1N-555, 'WAKEMASTER', c.1950

PLASTIC, 5 TUBES, 1 BAND

$55

Lafayette

1N-559, c.1950

WOOD, 5 TUBE, 1 BAND

$35

Lafayette

1N-562, c.1950

PLASTIC, 5 TUBE, 1 BAND

$45

Lafayette

1N-819, c.1950

WOOD, 8 TUBE, AM-FM

$40

Lafayette

25N-24305, Kit, c.1950
BAKELITE, 5 TUBES, 2 BANDS
$50

Lafayette

25N-24548, Kit, c.1950
BAKELITE, 5 TUBES, 1 BAND
$50

Lafayette

68B, c.1931
WOOD, 6 TUBES, 1 BAND
$225

Lafayette

115K, c.1953
BAKELITE, 5 TUBES, 1 BAND
$45

Lafayette

812K, c.1953
BAKELITE, 5 TUBES, 2 BANDS
$35

Lafayette

B40, c.1935
WOOD, 7 TUBES, 2 BANDS
$200

Lafayette

BB2, c.1940
WOOD, 8 TUBES, 3 BANDS
$95

Lafayette

BB27, c.1940
WOOD, 6 TUBES, 2 BANDS
$110

Lafayette

BS3, c.1940
WOOD, 8 TUBES, 3 BANDS
$125

Lafayette

C36, 'MONARCH', c.1931
WOOD, 6 TUBES, 1 BAND
$225

Lafayette

C54(BROWN), C60(BLACK),
C66(IVORY), c.1938
BAKELITE/PLASKON, 6 TUBES, 1 BAND
BROWN $225, BLACK $275
IVORY $350

Lafayette

C69, RADIO-PHONO, c.1938
WOOD, 9 TUBES, 3 BANDS
$75

Lafayette

C77, c.1938

WOOD, 12 TUBES, 4 BANDS

$175

Lafayette

C200, c.1942

WOOD, 8 TUBES, 1 BAND

$110

Lafayette

C205, c.1942

WOOD, 8 TUBES, 3 BAND

$125

Lafayette

C219, c.1942

WOOD, 6 TUBES, 1 BAND

$50

Lafayette

C251, c.1942

WOOD, 7 TUBES, 2 BANDS

$65

Lafayette

CC1, c.1940

WOOD, 7 TUBES, 3 BANDS, DC

$45

Lafayette

CC47, c.1940

WOOD, 12 TUBES, 3 BANDS

$165

Lafayette

CC57T, c.1940

WOOD, 9 TUBES, 3 BANDS, DC

$135

Lafayette

CC84, c.1940

WOOD, 5 TUBES, 1 BAND

$75

Lafayette

D19, c.1937

WOOD, 6 TUBES, 3 BANDS

$90

Lafayette

D26, c.1940

WOOD, 6 TUBES, 2 BANDS

$80

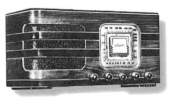

Lafayette

D32, c.1937

WOOD, 6 TUBES, 2 BANDS

$85

Lafayette

D38, c.1937

WOOD, 5 TUBES, 1 BAND

$90

Lafayette

D42, c.1940

WOOD, 7 TUBES, 3 BANDS

$115

Lafayette

D66, c.1940

WOOD, 6 TUBES, 1 BAND

$60

Lafayette

D68, c.1938

WOOD, 7 TUBES, 2 BANDS

$125

Lafayette

D69G, c.1940

WOOD, 8 TUBES, 1 BAND

$90

Lafayette

D71G, c.1940

WOOD, 8 TUBES, 3 BANDS

$125

Lafayette

D247, c.1942

WOOD, 5 TUBES, 1 BAND

$65

Lafayette

D291, c.1942

WOOD, 5 TUBES, 1 BAND

$45

Lafayette

D294, c.1942

WOOD, 6 TUBES, 2 BAND

$45

Lafayette

DA14, c.1938

WOOD, 4 TUBES, 2 BANDS, DC

$40

Lafayette

DA16, c.1937

WOOD, 7 TUBES, 3 BANDS, DC

$50

Lafayette

DA20, c.1937

WOOD, 6 TUBES, 3 BANDS

$90

Lafayette

DA28, c.1937
WOOD, 7 TUBES, 3 BANDS, DC
$50

Lafayette

DA31, c.1937
WOOD, 7 TUBES, 3 BANDS
$125

Lafayette

E20, c.1940
WOOD, 5 TUBES, 2 BANDS, DC
$25

Lafayette

E22, c.1940
BAKELITE, 5 TUBES, 2 BANDS
$75

Lafayette

E29, c.1940
WOOD, 5 TUBES, 2 BANDS
$115

Lafayette

E38, c.1940
BAKELITE, 5 TUBES, 1 BAND
$50

Lafayette

E59, c.1940
WOOD, 5 TUBES, 2 BANDS, DC
$25

Lafayette

EM, c.1931
WOOD, 8 TUBES, 1 BAND
$225

Lafayette

F36, c.1935
WOOD
$225

Lafayette

F57, c.1935
WOOD, 10 TUBES, 4 BANDS, DC
$150

Lafayette

FE6, c.1940
BAKELITE, 6 TUBES, 1 BAND
$135

Lafayette

FE28, c.1940
BAKELITE, 7 TUBES, 2 BANDS
$55

Lafayette

FE1426, c.1942
BAKELITE, 5 TUBES, 1 BAND
$245

Lafayette

FM1, FM CONVERTER, c.1940
WOOD, FM ONLY
$65

Lafayette

FS11, RADIO-PHONO, c.1938
WOOD, 5 TUBES, 1 BAND
$60

Lafayette

FS17, c.1938
WOOD, 5 TUBES, 1 BAND
$70

Lafayette

FS45, RADIO-PHONO, c.1937
WOOD, 5 TUBES, 1 BAND
$70

Lafayette

FS73, c.1938
WOOD, 6 TUBES, 2 BANDS, DC
$60

Lafayette

Y5B(BROWN), Y5I(IVORY), c.1948
BAKELITE/PLASKON
5 TUBES, 1 BAND
BROWN $85, IVORY $150

Lafayette

JA35, c.1937
WOOD, 5 TUBES, 1 BAND
$60

Lafayette

JA328, c.1942
WOOD, 5 TUBES, 1 BAND
$65

Lafayette

JB7, c.1937
WOOD, 5 TUBES, 1 BAND
$115

Lafayette

JL5, JL5Y, c.1948
BAKELITE, 5 TUBES, 1 BAND
$35

Lafayette

JL7, c.1948
WOOD, 7 TUBES, 2 BANDS
$45

Lafayette
JS187, c.1942
BAKELITE/PLASKON, 6 TUBES, 2 BANDS
BROWN $100, IVORY $150

Lafayette
JS189, c.1942
BAKELITE/PLASKON, 6 TUBES, 3 BANDS
BROWN $100, IVORY $150

Lafayette
JS190, c.1942
WOOD, 6 TUBES, 3 BANDS
$65

Lafayette
K20360, FM CONV., c.1948
WOOD, 5 TUBES, FM ONLY
$35

Lafayette
LE5, c.1953
BAKELITE, 4 TUBES, 1 BAND
$75

Lafayette
M19, c.1948
WOOD, 5 TUBES, AM/FM
$60

Lafayette
MA3, c.1942
WOOD, 6 TUBES, 3 BANDS
$75

Lafayette
MA3, c.1940
WOOD, 7 TUBES, 4 BANDS, DC
$50

Lafayette
MA322, c.1942
WOOD, 6 TUBES, 1 BAND
$40

Lafayette
MC16B(BROWN), MC16Y(IVORY), c.1948
WOOD, 4 TUBES, 1 BAND
$75

Lafayette
"NOMAD", c.1933
WOOD, 5 TUBES, 1 BAND
$225

Lafayette
P, "GEM", c.1931
WOOD, 6 TUBES, 1 BAND
$150

Lafayette

S165, c.1942
WOOD, 5 TUBES, 1 BAND
$55

Lafayette

SW CONVERTER, c.1931
METAL, 3 TUBES, SW
$75

Lafayette

T99, c.1940
BAKELITE, 4 TUBES, 1 BAND
$75

Lafayette

"WIDE WORLD", c.1931
METAL, 6 TUBES, SW
$100

Lang

50AS, c.1935
WOOD, 5 TUBES, 2 BANDS
$180

LaSalle

"C", c.1935
WOOD & ALUMINUM
4 TUBES, 1 BAND
$325

LaSalle

"F", c.1935
WOOD, 5 TUBES, 2 BANDS
$175

LaSalle

"J", "LEADER", c.1935
WOOD, 4 TUBES, 1 BAND
$295

LaSalle

"K", c.1935
WOOD, 4 TUBES, 2 BANDS
$375

LaSalle

"L", c.1935
WOOD, 5 TUBES, 2 BANDS
$475

LaSalle

"M", c.1935
WOOD, 4 TUBES, 1 BAND
$75

LaSalle

"R", c.1935
WOOD, 5 TUBES, 1 BAND
$135

LaSalle

SW CONVERTER, C.1935
WOOD, 2 TUBES
$65

LITTLE GIANT

"LITTLE GIANT", C.1934
WOOD, CRYSTAL
$500

Loftin-White

SCREEN GRID, C.1931
METAL
$115

Lormel

KIT, C.1951
MARBELED PLASTIC, 4 TUBES, 1 BAND
$150

LYRIC

U50, C.1934
WOOD, 5 TUBES, 1 BAND
$175

Majestic

51(BROWN), 52(IVORY), C.1947
BAKELITE/PAINTED, 5 TUBES, 1 BAND
$115

Majestic

55B, C.1934
WOOD, 5 TUBES, 1 BAND
$750

Majestic

105, C.1934
WOOD, 5 TUBES, 1 BAND
$250

Majestic

151, "HAVENWOOD", C.1932
WOOD
$135

Majestic

413, "KNOCKABOUT", C.1933
WOOD, 6 TUBES, 1 BAND
$90

Majestic

T202A, C.1941
WOOD
$65

Mantola

C.1938
WOOD
$75

MARCONI

c.1934

WOOD, 8 TUBES, 2 BANDS

$650

MARCONI

4T2, c.1959

PLASTIC, 4 TUBES, 1 BAND

$40

MARCONI

60DC, c.1936

WOOD, 5 TUBES, 1 BAND, DC

$70

MARCONI

62AC, c.1936

WOOD, 5 TUBES, 1 BAND

$225

MARCONI

64DC, c.1936

WOOD, 6 TUBES, 2 BANDS, DC

$125

MARCONI

66AC, c.1936

WOOD, 7 TUBES, 2 BANDS

$275

MARCONI

73AC, c.1936

WOOD, 7 TUBES, 1 BAND

$200

MARCONI

153, c.1940

BAKELITE, 5 TUBES, 1 BAND

$75

MARCONI

164, RADIO-PHONO, c.1940

WOOD, 5 TUBES, 1 BAND

$50

MARCONI

166, c.1940

WOOD, 5 TUBES, 2 BANDS

$60

MARCONI

167, c.1940

WOOD, 6 TUBES, 5 BANDS

$85

MARCONI

169, c.1940

WOOD, 8 TUBES, 3 BANDS

$90

MARCONI

216, C.1941

BAKELITE, 5 TUBES, 1 BAND

$45

MARCONI

217A, RADIO-PHONO, C.1941

WOOD, 5 TUBES, 1 BAND

$35

MARCONI

217B, RADIO-PHONO, C.1941

WOOD, 5 TUBES, 1 BAND

$35

MARCONI

217SW, C.1941

WOOD, 5 TUBES, 2 BANDS

$45

MARCONI

218, C.1941

BAKELITE, 5 TUBES, 1 BAND

$75

MARCONI

219, C.1941

WOOD, 5 TUBES, 2 BANDS

$65

MARCONI

221, RADIO-PHONO, C.1941

WOOD, 5 TUBES, 2 BANDS

$65

MARCONI

222, RADIO-PHONO, C.1941

WOOD, 5 TUBES, 2 BANDS

$65

MARCONI

227, C.1941

BAKELITE, 4 TUBES, 1 BAND, DC

$35

MARCONI

227A, C.1941

WOOD, 4 TUBES, 1 BAND, DC

$25

MARCONI

228, C.1941

WOOD, 5 TUBES, 2 BANDS, DC

$25

MARCONI

230, C.1941

BAKELITE, 4 TUBES, 1 BAND, DC

$25

MARCONI

231, C.1941

BAKELITE, 5 TUBES, 1 BAND, DC

$25

MARCONI

233, C.1941

BAKELITE, 7 TUBES, 3 BANDS, DC

$25

MARCONI

236, C.1941

BAKELITE, 5 TUBES, 1 BAND

$35

MARCONI

238, C.1949

WOOD, 6 TUBES, 2 BANDS

$35

MARCONI

252FM, C.1949

WOOD, 14 TUBES, 2 BAND + FM

$65

MARCONI

254FM, C.1949

WOOD, 9 TUBES, AM-FM

$45

MARCONI

258, C.1949

BAKELITE, 5 TUBES, 1 BAND

$65

MARCONI

261, C.1948

BAKELITE & METAL
5 TUBES, 1 BAND

$50

MARCONI

264, C.1948

BAKELITE, 5 TUBES, 1 BAND

$50

MARCONI

267, C.1948

WOOD, 5 TUBES, 2 BANDS, DC

$30

MARCONI

271, C.1949

BAKELITE, 5 TUBES, 1 BAND

$45

MARCONI

271A, C.1949

BAKELITE, 5 TUBES, 1 BAND

$45

MARCONI

274, RADIO-PHONO, C.1948
WOOD, 5 TUBES, 1 BAND
$45

MARCONI

275, C.1948
BAKELITE, 5 TUBES, 1 BAND
$35

MARCONI

275A, C.1948
WOOD, 5 TUBES, 1 BAND
$30

MARCONI

276, C.1958
BAKELITE, 9 TUBES, AM-FM
$45

MARCONI

279, C.1948
WOOD, 6 TUBES, 1 BAND
$30

MARCONI

288, C.1948
BAKELITE, 5 TUBES, 1 BAND
$60

MARCONI

289, C.1950
BAKELITE, 5 TUBES, 1 BAND
$45

MARCONI

290, C.1950
BAKELITE, 4 TUBES, 1 BAND
$45

MARCONI

291, C.1950
BAKELITE, 5 TUBES, 1 BAND
$25

MARCONI

294, C.1950
BAKELITE, 5 TUBES, 1 BAND
$60

MARCONI

302, 303, C.1951
BAKELITE, 5 TUBES, 1 BAND
$25

MARCONI

305, C.1950
BAKELITE, 5 TUBES, 1 BAND
$40

MARCONI

317, c.1951

BAKELITE, 4 TUBES, 1 BAND

$50

MARCONI

339, c.1952

BAKELITE, 5 TUBES, 1 BAND

$30

MARCONI

341, c.1955

BAKELITE, 6 TUBES, 1 BAND

$30

MARCONI

342, c.1954

BAKELITE, 5 TUBES, 1 BAND

$35

MARCONI

349, c.1955

PLASTIC, 5 TUBES, 1 BAND

$35

MARCONI

355, c.1952

BAKELITE, 5 TUBES, 1 BAND

$40

MARCONI

359, c.1956

PLASTIC, 5 TUBES, 1 BAND

$35

MARCONI

362, c.1953

PLASTIC, 4 TUBES, 1 BAND

$40

MARCONI

379, c.1956

PLASTIC, 6 TUBES, 1 BAND

$25

MARCONI

385, c.1955

BAKELITE, 5 TUBES, 1 BAND

$35

MARCONI

389, c.1955

WOOD, 6 TUBES, 1 BAND

$25

MARCONI

395, c.1957

WOOD, 8 TUBES, 3 BANDS

$30

MARCONI

399, c.1955

WOOD, 8 TUBES, 2 BANDS

$30

MARCONI

406, c.1955

BAKELITE, 5 TUBES, 1 BAND

$30

MARCONI

407, 'MIGHTY MITE', c.1955

BAKELITE, 5 TUBES, 1 BAND

$45

MARCONI

409, c.1956

WOOD, 6 TUBES, 1 BAND

$25

MARCONI

417, c.1959

PLASTIC, 5 TUBES, 1 BAND

$30

MARCONI

422, c.1955

PLASTIC, 5 TUBES, 1 BAND

$45

MARCONI

425, c.1959

PLASTIC, 5 TUBES, 1 BAND

$45

MARCONI

439, c.1959

WOOD, 6 TUBES, 1 BAND

$25

MARCONI

462, c.1959

PLASTIC, 5 TUBES, 1 BAND

$55

MARCONI

472, c.1959

PLASTIC, 5 TUBES, 1 BAND

$55

MARCONI

482, c.1959

PLASTIC, 5 TUBES, 1 BAND

$55

MARCONI

588, c.1959

WOOD, TRANSISTOR

$20

MARCONI

623, c.1959

PLASTIC, 6 TUBES, 1 BAND

$30

MARCONI

1003, 'PLAYBOY', c.1948

WOOD, PHONO ONLY, 2 TUBES

$25

MARCONI

1005, 1006, 1007, c.1948

WOOD, PHONO REMOTE

$25

MARCONI

'CLASSIC', c. 1937

METAL, 4 TUBES, 1 BAND, DC

$500+

MARCONI

'JUNIOR', c.1947

WOOD, PHONO REMOTE

$25

MARVEL

SW CONVERTER, c.1936

BAKELITE, 1 TUBE, SHORTWAVE

$225

MARVELO

MIDGET, c.1931

WOOD, 6 TUBES, 1 BAND

$250

MASCOT

TREASURE CHEST, c.1933

WOOD, 5 TUBES, 1 BAND

$225

MASCOT

COMRADE, c.1933

WOOD, 5 TUBES, 1 BAND

$125

MASCOT

DUAL WAVE, c.1933

WOOD, 6 TUBES, 2 BANDS

$125

MASTER

50, c.1931

WOOD

$75

MAY

T35, c.1935

WOOD, 5 TUBES, 1 BAND

$165

McMurdo Silver

5D, c.1936
CHROME CHASSIS
$500+

McMurdo Silver

14-15, c.1938
CHROME CHASSIS
$750+

MECK

237, c.1946
BAKELITE, 5 TUBES, 1 BAND
$90

MECK

CC500, PHONO, c.1947
WOOD, 2 TUBES, PHONO ONLY
$25

MECK

CD500, RADIO-PHONO,
c.1947
WOOD, 5 TUBES, 1 BAND

MECK

CE500, c.1948
PLASTIC, 4 TUBES, 1 BAND
$65

MECK

CH500, c.1948
PLASTIC, 5 TUBES, 1 BAND
$30

MECK

CJ500, c.1948
PLASTIC, 5 TUBES, 1 BAND
$35

MECK

CM500, c.1948
PLASTIC, 5 TUBES, 1 BAND
$40

MECK

CP500, c.1948
PLASTIC, 4 TUBES, 1 BAND
$85

MECK

CR500, c.1947
WOOD, 10 TUBES, AMFM
$35

MEISSNER

1110, c.1940
KIT, 7 TUBES, 3 BANDS
$65

MEISSNER

1155, c.1940
KIT, 12 TUBES, 5 BANDS
$150

MEISSNER

1170, c.1940
KIT, 9 TUBES, 5 BANDS
$125

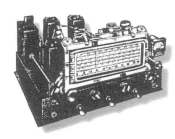

MEISSNER

1174, c.1940
KIT, 14 TUBES, 5 BANDS
$160

MEISSNER

MIDGET, c.1940
KIT, 1, 2 OR 3 TUBES, 1 BAND
$75

MEISSNER

REMOTE, c.1930
BAKELITE, 3 TUBES
$75

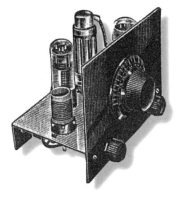

MEISSNER

MIDGET, c.1942
KIT, 1, 2 OR 3 TUBES, 1 BAND
$75

Melburn

c.1935
WOOD, 5 TUBES, 1 BAND
$90

MIDWEST

G-11, c.1936
WOOD, 11 TUBES, 4 BANDS
$850

MIDWEST

H-6, c.1934
WOOD, 6 TUBES, 2 BANDS
$450

MIDWEST

K-6, c.1960
WOOD, 6 TUBES, 2 BANDS
$250

MIDWEST

L-15, MERCURY, c.1940
WOOD, 15 TUBES, 2 BANDS
$350

MIDWEST

L-17, MERCURY, c.1940
WOOD, 17 TUBES, 5 BANDS
$500
ALSO AVAILABLE AS L-12 (12 TUBES, 5
BANDS) VALUED AT $400

MIDWEST

M-9, c.1940
WOOD, 9 TUBES, 3 BANDS
$300

MIDWEST

PH-5, RADIO-PHONO, c.1940
WOOD, 5 TUBES, 1 BAND
$100

MIDWEST

S-4, c.1940
BAKELITE, 4 TUBES, 1 BAND, DC
$50

Mission-Bell

375, c.1937
WOOD, 5 TUBES, 3 BANDS
$125

Mission-Bell

376, c.1937
WOOD, 5 TUBES, 1 BAND
$75

Mission-Bell

22, 'JACKSON BELL', c.1936
WOOD, 5 TUBES, 1 BAND
$350

MOHAWK

C, 'BLACKHAWK', c.1931
WOOD, 7 TUBES, 1 BAND
$325

Monarch

50, c.1938
WOOD, 5 TUBES, 1 BAND
$65

Monarch

5200, c.1933
WOOD, 10 TUBES, 1 BAND
$350

MONTERREY

c.1934
WOOD, 5 TUBES, 1 BAND
$400

Motorola

3A5, 'PLAYBOY', c.1942
METAL, 5 TUBES, 1 BAND
$50

Motorola

3F22, PHONO, c.1958
FIBERGLASS, 3 TUBES, PHONO ONLY
$75
AVAILABLE IN BLUE, RED OR BLACK

Motorola

3H24, c.1958
FIBERGLASS, 3 TUBES, PHONO
$75
AVAILABLE IN TAN OR BLUE

Motorola

41F, c.1940
WOOD, 4 TUBES, 1 BAND
$30

Motorola

5A9, c.1948
METAL & PLASTIC
4 TUBES, 1 BAND
$60

Motorola

5C13, c.1959
UREA, 5 TUBES, 1 BAND
5C13B(BLUE) $175,
5C13M(BROWN) $45
5C13P(PINK) $175
5C13W(WHITE) $50

Motorola

5C14, c.1959
PLASTIC, 5 TUBES, 1 BAND
5C14CW(BLUE) $90
5C14GW(GREEN) $90
5C14PW(PINK) $90

Motorola

5C15, c.1959
PLASTIC, 5 TUBES, 1 BAND
5C14BW(BLUE) $90
5C14GW(GREEN) $90
5C14VW(LAVENDER) $175

Motorola

5C16, c.1959
PLASTIC, 5 TUBES, 1 BAND
5C14NW(BROWN) $50
5C14W(WHITE) $60

Motorola

5C21, c.1959
UREA, 5 TUBES, 1 BAND
5C21B(BLUE) $175
5C21G(GREEN) $175
5C21P(PINK) $145

Motorola

5C22, c.1958
UREA, 5 TUBES, 1 BAND
BROWN $50, YELLOW $160
PINK $145, LAVENDER $250

Motorola

5C23, c.1958
PLASTIC, 5 TUBES, 1 BAND
BROWN $45, GREEN $90
PINK $90

Motorola

5C24, c.1958
PLASTIC, 5 TUBES, 1 BAND
BROWN $45, PINK $90

Motorola

5C25, c.1958
PLASTIC, 5 TUBES, 1 BAND
BROWN $45, GREEN $90

Motorola

5C27, c.1958
UREA, 5 TUBES, 1 BAND
BROWN $45, IVORY $90
VIOLET $245

Motorola

5R23, RADIO-PHONO, c.1958
FIBERGLASS, 5 TUBES, 1 BAND
$80

AVAILABLE IN BROWN, IVORY OR GREEN

Motorola

5T12, c.1958
PLASTIC, 5 TUBES, 1 BAND
5T12G(GREEN) $65
5T12M(BROWN) $35
5T12R(RED) $75
T12W(WHITE) $40

Motorola

5T13, c.1958
PLASTIC, 5 TUBES, 1 BAND
5T12P(TAN) $45
5T12S(BLACK) $40

Motorola

5T21, c.1958
PLASTIC
$45

Motorola

5T22, c.1958
PLASTIC, 5 TUBES, 1 BAND
BROWN $95, IVORY $110
RED $145, YELLOW $145

Motorola

5T23, c.1958
PLASTIC, 5 TUBES, 1 BAND
BROWN $35, IVORY $40
PINK $60, YELLOW $65

Motorola

5T25, c.1958
WOOD, 5 TUBES, 1 BAND
$35

Motorola

5T27, c.1958
UREA, 5 TUBES, 1 BAND
BROWN $75, IVORY $110
VIOLET $275

Motorola

5C26, c.1958
UREA, 6 TUBES, 1 BAND
GREY $65, IVORY $60

Motorola

6T26, c.1958
UREA, 6 TUBES, 1 BAND
BROWN $45, IVORY $50

Motorola

6X11(WALNUT), 6X12(IVORY)
c.1950
BAKELITE, 6 TUBES, 1 BAND
$75

Motorola

7XM21, c.1950
BAKELITE, 8 TUBES, AMFM
$75

Motorola

1OT28, c.1958
BAKELITE, 9 TUBES, AMFM
$35

Motorola

11A, c.1940
PHONO REMOTE
$25

Motorola

41F, c.1940
WOOD, 4 TUBES, 1 BAND, DC
$25

Motorola

42B, c.1952
METAL & PLASTIC
$50

Motorola

48L11, c.1948
PLASTIC, 4 TUBES, 1 BAND
$75

Motorola

49L11, c.1948
PLASTIC, 4 TUBES, 1 BAND, DC
$65

Motorola

51F11, 'RADIO-PHONO, c.1942
WOOD, 5 TUBES, 1 BAND
$55

Motorola

51X13(WALNUT),
51X14(IVORY), c.1942
BAKELITE/PLASKON
5 TUBES, 1 BAND
BROWN $45, IVORY $65

Motorola

51X17, c.1942
CANVAS COVERED WOOD
5 TUBES, 1 BAND
$35

Motorola

52C, c.1952
BAKELITE, 5 TUBES, 1 BAND
$40

Motorola

52CW, c.1952
PLASTIC, 5 TUBES, 1 BAND
$50

Motorola

52L, c.1952
FAUX LEATHER
$25

Motorola

52M, c.1952
PLASTIC & METAL
$60

Motorola

53C1(BROWN), 56C2(TAN),
56C3(GREEN), 56C4(BLUE),
c.1953
BAKELITE, 5 TUBES, 1 BAND
BROWN, TAN $50
BLUE, GREEN $70

Motorola

53C6(BROWN), 56C7(IVORY),
56C8(GREEN), 56C9(TAN),
c.1953
PLASTIC, 5 TUBES, 1 BAND
BROWN, IVORY, TAN $30
GREEN $50

Motorola

53D1, c.1953
METAL, 5 TUBES, 1 BAND
$125

Motorola

53R1(BROWN), 53R2(IVORY),
53R3(YELLOW), 53R4(GREY),
53R5(GREEN), 53R4(RED),
c.1953
BAKELITE, 5 TUBES, 1 BAND
BROWN, IVORY, GREY $45
YELLOW, GREEN RED $70

Motorola

53X1(BROWN), 53X2(TAN),
53X3(GREEN), 53X4(ORANGE),
c.1953
BAKELITE, 5 TUBES, 1 BAND
$95

Motorola

54X1(BROWN), 54X2(IVORY),
54X3(GREEN), c.1953
PLASTIC, 5 TUBES, 1 BAND
BROWN $35, IVORY $40,
GREEN $85

Motorola

55A1(BROWN), 55A2(BLACK),
55A3(IVORY), c.1955
PLASTIC, 5 TUBES, 1 BAND
BROWN $35, BLACK $40,
IVORY $40

Motorola

56CJ1(BLACK), 56CJ2(IVORY),
c.1956
PLASTIC, 5 TUBES, 1 BAND
BLACK $50, IVORY $50

Motorola

56W, c.1956
WOOD, 5 TUBES, 1 BAND
$25

Motorola

57A1(BLACK), 57A2(RED),
57A3(IVORY), c.1957
PLASTIC, 5 TUBES, 1 BAND
BLACK $45, RED $90, IVORY
$45

Motorola

57CD1(BROWN), 57CD2(IVORY),
57CD3(BLUE), 57CD4(AQUA),
C.1955
UREA, 5 TUBES, 1 BAND
BROWN $65, IVORY $80
BLUE $175, AQUA $195

Motorola

57R1(BLACK), 57R2(IVORY),
57R3(AQUA), 57R4(PURPLE), C.1955
UREA, 5 TUBES, 1 BAND
BLACK $90, IVORY $90
AQUA $175, PURPLE $275

Motorola

58A11(BROWN), 58A12(IVORY),
C.1949
BAKELITE/PLASKON, 5 TUBES, 1 BAND
BROWN $60, IVORY $90

Motorola

58R11(BROWN), 58R12(BLACK),
58R13(GREEN), 58R14(YELLOW),
58R15(TAN), 58R16(IVORY), C.1949
BAKELITE/PAINTED, 4 TUBES, 1 BAND
$95

Motorola

'PYRAMID', C.1949
58X11(BROWN), 58X12(IVORY)
BAKELITE/PAINTED, 5 TUBES, 1 BAND
$60

Motorola

59L12, C.1948
PLASTIC, 4 TUBES, 1 BAND
$60

Motorola

59T1, C.1939
WOOD, 5 TUBES, 1 BAND
$90

Motorola

59T2, C.1939
WOOD, 5 TUBES, 1 BAND
$95

Motorola

59T3, C.1939
WOOD, 5 TUBES, 1 BAND
$95

Motorola

61F, RADIO-PHONO, C.1940
WOOD, 6 TUBES, 1 BAND
$60

Motorola

61L11, 'PLAYMATE', C.1942
6 TUBES, 1 BAND
$45

Motorola

62C, C.1952
PLASTIC, 5 TUBES, 1 BAND
$65

Motorola

62CW1, c.1953
WOOD, 6 TUBES, 1 BAND
$40

Motorola

62L1(GREEN), 62L2(MAROON),
62L3(GREY), c.1952
PLASTIC, 5 TUBES, 1 BAND
GREEN $75, MAROON $75, GREY $50

Motorola

62X11(BROWN), 62X12(IVORY),
62X13(GREEN), c.1952
BAKELITE/PAINTED, 6 TUBES, 1 BAND
$65

Motorola

63X1(BLACK), 63X1A(BROWN),
63X2(IVORY), 63X3(GREEN), c.1953
PLASTIC, 6 TUBES, 1 BAND
BLACK $35, BROWN $25
IVORY $35, GREEN $65

Motorola

65T21B, c.1946
WOOD, 6 TUBES, AMFM
$35

Motorola

66C1(IVORY), 66C2(GREY)
c.1956
PLASTIC, 6 TUBES, 1 BAND
$50

Motorola

68F11, RADIO-PHONO, c.1949
BAKELITE, 5 TUBES, 1 BAND
$65

Motorola

68F12, RADIO-PHONO, c.1949
WOOD, 5 TUBES, 1 BAND
$65

Motorola

68T11, c.1949
BAKELITE, 5 TUBES, 1 BAND
$65

Motorola

68T12, c.1949
WOOD, 5 TUBES, 1 BAND
$35

Motorola

68X11(MAROON),
68X12(BROWN), c.1949
PLASTIC, 5 TUBES, 1 BAND
MAROON $95, BROWN $75

Motorola

69L11, c.1948

$75

Motorola

71A, c.1940

WOOD, 7 TUBES, 3 BANDS

$125

Motorola

77XM22, c.1949

WOOD, 5 TUBES, 1 BAND

$60

Motorola

496BT1, c.1940

WOOD, 4 TUBES, 1 BAND, DC

$25

Motorola

A7, c.1960

PLASTIC, 4 TUBES, 1 BAND

$45

Motorola

A9, c.1960

PLASTIC, 5 TUBES, 1 BAND

$45

Motorola

A12, c.1960

PLASTIC, 7 TUBES, 1 BAND

$75

Motorola

A15, c.1961

PLASTIC, 5 TUBES, 1 BAND

$45

Motorola

A23, c.1961

PLASTIC, 5 TUBES, 1 BAND

$35

Motorola

A24, c.1962

PLASTIC, 5 TUBES, 1 BAND

$75

Motorola

AX5, c.1961

PLASTIC, TRANSISTOR

$40

Motorola

B1, c.1960

PLASTIC, AMFM

$65

Motorola

B2, c.1960

PLASTIC, AMFM

$65

Motorola

B3, c.1960
PLASTIC, AMFM
$65

Motorola

B5, c.1960
PLASTIC, AMFM
$65

Motorola

BC1, c.1961
PLASTIC, AMFM
$35

Motorola

BC2(TAN), BC3(GREEN),c.1961
PLASTIC, 8 TUBES, AMFM
TAN $35, GREEN $45

Motorola

C11, c.1960
PLASTIC, 5 TUBES, 1 BAND
$65

Motorola

C12, c.1960
PLASTIC, 6 TUBES, 1 BAND
$45

Motorola

C15, c.1960
PLASTIC, 5 TUBES, 1 BAND
$45

Motorola

C26, c.1960
PLASTIC, 5 TUBES, 1 BAND
$25

Motorola

CX28, c.1961
PLASTIC, TRANSISTOR, 1 BAND
$45

Motorola

GN, c.1958
PLASTIC, 5 TUBES, 1 BAND
$35

Motorola

'VITA-TONE', c.1942
WOOD, 6 TUBES, 1 BAND
$45

Motorola

XT18, c.1960
PLASTIC, TRANSISTOR, 1 BAND
$25

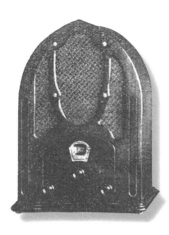

Musique

MUSIQUE 5, C.1932
WOOD, 5 TUBES, 1 BAND
$150

Musique

MUSIQUE 8, C.1932
WOOD, 8 TUBES, 1 BAND
$225

NAVIGATOR

LONGWAVE, C.1932
WOOD, 6 TUBES, 1 BAND
$750

NAVIGATOR

SUPER 6, C.1932
WOOD, 6 TUBES, 1 BAND
$225

NAVIGATOR

SUPER 8, C.1932
WOOD, 8 TUBES, 1 BAND
$750

New Era

SHORTWAVE, C.1934
WOOD, 2 TUBES, 1 BAND
$110

New Marvel

CRYSTAL, C.1934
BAKELITE, CRYSTAL, 1 BAND
$175

Newcomb

B-100, C.1951
WOOD
$20

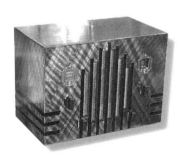

Northern Electric

53, C.1934
WOOD, 5 TUBES, 1 BAND
$375

Northern Electric

518, C.1957
PLASTIC, 5 TUBES, 1 BAND
$40

Northern Electric

548, C.1957
PLASTIC, 5 TUBES, 1 BAND
$40

Northern Electric

598, C.1957
PLASTIC, 5 TUBES, 1 BAND
$40

Northern **NE** *Electric*

607, C.1957
PLASTIC, 5 TUBES, 1 BAND
$50

Northern **NE** *Electric*

616, C.1958
PLASTIC, TRANSITOR, 1 BAND
$175

Northern **NE** *Electric*

1101, C.1957
PLASTIC, 5 TUBES, 1 BAND
$30

Northern **NE** *Electric*

1102, C.1957
PLASTIC, 5 TUBES, 1 BAND
$45

Northern **NE** *Electric*

1102 EXTENSION, C.1957
PLASTIC, 5 TUBES, 1 BAND
$40

Northern **NE** *Electric*

1201, C.1957
PLASTIC, 5 TUBES, 1 BAND
$40

Northern **NE** *Electric*

2302, C.1957
PLASTIC, 5 TUBES, 1 BAND
$40

Northern **NE** *Electric*

5410, C.1958
PLASTIC, 4 TUBES, 1 BAND
$80

Northern **NE** *Electric*

5412 'SKY CHAMP', C.1958
PLASTIC, 5 TUBES, 1 BAND
$35

Northern **NE** *Electric*

5500 'BABY CHAMP', C.1958
PLASTIC, 5 TUBES, 1 BAND
$45

Northern **NE** *Electric*

5508 'BULLET', C.1958
PLASTIC, 5 TUBES, 1 BAND
$150

Northern **NE** *Electric*

5700 'BABY CHAMP', C.1953
PLASTIC, 5 TUBES, 1 BAND
$45

Northern **N/E** *Electric*

5708, C.1954
BAKELITE, 5 TUBES, 1 BAND
$75

Oriole

3-6B, MOTOR TUNING, C.1938
WOOD, 6 TUBES, 2 BANDS
$125

Oriole

3-A2I (IVORY), 3-A2B (BLACK), C.1938
PLASKON/BAKELITE
7 TUBES, 3 BANDS
IVORY $225, BLACK $145

Oriole

3C-130, C.1940
WOOD, 6 TUBES, 2 BANDS
$85

Oriole

3F-56, C.1937
WOOD, 5 TUBES, 2 BANDS
$110

Oriole

3W-200, C.1940
WOOD, 8 TUBES, 3 BANDS
$95

Oriole

3W-201, C.1940
WOOD, 7 TUBES, 3 BANDS
$95

Oriole

3X-100, C.1937
WOOD, 4 TUBES, 1 BANDS
$90

Oriole

4F-101, C.1940
WOOD, 6 TUBES, 1 BAND
$50

Oriole

30C-66, C.1953
BAKELITE, 5 TUBES, 1 BAND
$35

Oriole

36B-17, C.1956
BAKELITE, 5 TUBES, 1 BAND
$40

Oriole

36B-18, C.1956
MARBELED PLASTIC
6 TUBES, 1 BAND
$115

36B-19, c.1956

PLASTIC, 6 TUBES, 1 BAND

$40

102, c.1939

WOOD, 5 TUBES, 1 BAND

$60

502-6B, MOTOR TUNING, c.1938

WOOD, 6 TUBES, 2 BANDS

$100

512-6D, c.1938

WOOD, 6 TUBES, 2 BANDS, DC

$30

513-5A, c.1938

BAKELITE, 5 TUBES, 1 BAND

$225

516-5C, c.1938

BAKELITE, 5 TUBES, 1 BAND

$75

3112, c.1942

WOOD, 4 TUBES, 1 BAND, DC

$30

3113, c.1942

WOOD, 4 TUBES, 1 BAND, DC

$30

3199, RADIO-PHONO, c.1942

WOOD, 5 TUBES, 1 BAND

$50

4142, c.1942

WOOD, 17 TUBES, 2 BANDS + FM

$115

46048, c.1942

WOOD, 7 TUBES, 3 BANDS, DC

$75

46148, c.1942

WOOD, 7 TUBES, 3 BANDS, DC

$75

B3115, c.1940
WOOD, 4 TUBES, 1 BAND, DC
$40

C3-101, c.1940
BAKELITE, 5 TUBES, 1 BAND
$85

C3-105, c.1940
WOOD, 7 TUBES, 2 BANDS
$75

C3-113, c.1940
WOOD, 4 TUBES, 1 BAND, DC
$35

C3-137, c.1940
BAKELITE, 5 TUBES, 2 BANDS
$85

C3-155, c.1940
WOOD, 6 TUBES, 1 BAND
$60

DT-1595, c.1937
WOOD, 5 TUBES, 2 BANDS
$90

DT-1795, c.1937
WOOD, 6 TUBES, 2 BANDS
$145

DT-2595, c.1937
WOOD, 8 TUBES, 3 BANDS
$175

P12ENI, c.1938
WOOD, 6 TUBES, 3 BANDS
$115

PC945, c.1937
BAKELITE, 5 TUBES, 2 BANDS
$90

PC1365, c.1937
BAKELITE/PLASKON, 6 TUBES, 2 BANDS
BROWN $115, IVORY PLASKON $225
RED PLASKON $950, BEETLE $375

PC2145, c.1937
WOOD, 8 TUBES, 3 BANDS
$150

R3108, c.1940
WOOD, 5 TUBES, 2 BANDS, DC
$40

R3110, c.1940
WOOD, 4 TUBES, 1 BANDS, DC
$30

R3112, c.1940
WOOD, 4 TUBES, 1 BANDS, DC
$30

R4102, c.1940
BAKELITE/PLASKON, 6 TUBES, 2 BANDS
BROWN $135, IVORY $250

ST100, c.1937
WOOD, 6 TUBES, 2 BANDS, DC
$65

W3-100, c.1940
BAKELITE, 7 TUBES, 2 BANDS
$150

WA-1596, c.1937
WOOD, 5 TUBES, 2 BANDS, DC
$60

WA2500, c.1937
WOOD, 5 TUBES, 2 BANDS, DC
$45

WA3195, c.1937
WOOD, 8 TUBES, 3 BANDS
$275

Pacific

5E7, c.1937
WOOD, 5 TUBES, 1 BAND
$50

Pacific

15A1, c.1937
BAKELITE, 5 TUBES, 2 BANDS
$90

Pacific

15E4, c.1937

WOOD, 5 TUBES, 2 BANDS

$50

Pacific

20A2, c.1937

BAKELITE/PLASKON, 5 TUBES, 2 BANDS
BROWN $115, IVORY PLASKON $225
RED PLASKON $950, BEETLE $375

Pacific

20C5, c.1937

WOOD, 5 TUBES, 2 BANDS

$65

Pacific

30ME5, c.1937

WOOD, 7 TUBES, 2 BANDS

$65

Pacific

40MB1, c.1937

WOOD, 8 TUBES, 3 BANDS

$225

Pacific

40MC1, c.1937

WOOD, 8 TUBES, 3 BANDS

$225

Pacific

40ME1, c.1937

WOOD, 8 TUBES, 3 BANDS

$175

Pacific

50E6, c.1937

WOOD, 5 TUBES, 3 BANDS

$75

Pacific

51C3, c.1937

WOOD, 5 TUBES, 3 BANDS

$140

Pacific

60MB2, c.1937

WOOD, 7 TUBES, 3 BANDS

$175

Pacific

60MC2, c.1937

WOOD, 7 TUBES, 3 BANDS

$175

Pacific

3215, c.1937

WOOD, 5 TUBES, 1 BAND

$75

Packard-Bell

5R1, c.1956
PLASTIC, 5 TUBES, 1 BAND
$35

Packard-Bell

5T, c.1938
WOOD, 5 TUBES, 1 BAND
$75

Packard-Bell

35, c.1938
WOOD, 5 TUBES, 1 BAND
$115

Packard-Bell

46, "TWO-TONE", c.1938
WOOD, 6 TUBES, 2 BANDS
$175

Packard-Bell

46, c.1938
WOOD, 6 TUBES, 2 BANDS
$125

Packard-Bell

48, c.1938
WOOD, 8 TUBES, 3 BANDS
$175

Packard-Bell

50, c.1938
WOOD, 10 TUBES, 3 BANDS
$250

Packard-Bell

501, c.1938
PLASTIC, 5 TUBES, 1 BAND
$75

Packard-Bell

"KOMPAK", c.1938
MIRROR, 5 TUBES, 1 BAND
WHITE $800+, BLUE $1,100+

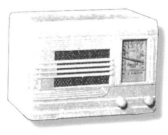

Packard-Bell

"KOMPAK", c.1938
BAKELITE/PLASKON, 5 TUBES, 1 BAND
BROWN $75, IVORY $150
RED $750, BLUE $750

Pandora

CRYSTAL, c.1935
BAKELITE, CRYSTAL, ONE BAND
$350

Paramount

5C, c.1936
WOOD, 5 TUBES, 2 BANDS
$110

Paramount

5S, c.1936

WOOD, 5 TUBES, 1 BAND

$90

Paramount

8J, c.1936

WOOD, 8 TUBES, 5 BANDS

$220

Paramount

369S, c.1936

WOOD, 6 TUBES, 2 BANDS

$110

Paramount

AA25, c.1936

WOOD, 4 TUBES, 1 BANDS

$350

Paramount

AB25, c.1936

WOOD, 5 TUBES, 1 BAND

$450

Paramount

E39B, c.1936

WOOD, 5 TUBES, 3 BANDS

$135

Paramount

E6T, c.1936

WOOD, 6 TUBES, 3 BANDS

$150

Paramount

G351, c.1936

WOOD, 5 TUBES, 1 BANDS

$70

Paramount

M35B, c.1936

WOOD, 6 TUBES, 3 BANDS

$175

Paramount

N351P, c.1936

WOOD, 5 TUBES, 2 BANDS

$60

Paramount

SC4, c.1936

WOOD, 4 TUBES, 1 BAND

$125

PARMAK

c.1937

WOOD, 6 TUBE, 2 BAND

$175

Pfanstiel

MIDGET, C.1932

WOOD, 5 TUBES, 1 BAND

$150

PHILCO

5B, C.1938

WOOD, 8 TUBES, 2 BANDS

$175

PHILCO

12C, C.1938

WOOD, 5 TUBES, 1 BAND

$65

PHILCO

12T, 14T, 15T, C.1938

5 TUBES, 1 BAND (12T),
2 BANDS (14T, 15T)

$65

PHILCO

14B, 16B, 18, C.1933

14B: 9 TUBES, 1 BAND
16B: 11 TUBES, 2 BANDS
18: 8 TUBES, 1 BAND

14B $250, 16B $350, 18 $225

PHILCO

22CL, C.1941

WOOD, 5 TUBES, 1 BAND

$150

PHILCO

28C, 45C,"BUTTERFLY", C.1935

WOOD, 6 TUBES, 2 BANDS

$300

PHILCO

29TX REMOTE, C.1935

WOOD, 6 TUBES, 2 BANDS

$275

PHILCO

29TX SPEAKER, C.1935

WOOD, SPEAKER ENCLOSURE

$200

PHILCO

39T, 40T, C.1938

39T: WOOD, 6 TUBES, 2 BANDS, DC
40T: WOOD, 6 TUBES, 2 BANDS

39T $55, 40T $125

PHILCO

43B, 60B, 89B, C.1933

14B: WOOD, 8 TUBES, 2 BANDS
16B: WOOD, 7 TUBES, 1 BAND
18: WOOD, 6 TUBES, 1 BAND

43B $300, 60B $275, 89B $250

PHILCO

54C, 57C, C.1933

WOOD, 5 TUBES, 1 BAND

$100

PHILCO

60B, 89B, c.1938
60B: WOOD, 5 TUBES, 1 BAND
89B: WOOD, 6 TUBES, 2 BANDS
60B $110, 89T $125

PHILCO

60B, c.1935
WOOD, 5 TUBES, 2 BANDS
$175

PHILCO

60MB, c.1934
WOOD, 5 TUBES, 1 BAND
$160

PHILCO

62T, c.1938
WOOD, 5 TUBES, 2 BANDS
$75

PHILCO

84B, c.1936
WOOD, 4 TUBES, 1 BAND
$110

PHILCO

93B, c.1938
WOOD, 5 TUBES, 2 BANDS
$80

PHILCO

100, c.1949
BAKELITE, 4 TUBES, 1 BAND
$40

PHILCO

101, c.1949
BAKELITE, 4 TUBES, 1 BAND
$35

PHILCO

118B, 49B, c.1938
118B: WOOD, 8 TUBES, 2 BANDS
49B: WOOD, 8 TUBES, 2 BANDS, DC
118B $110, 89T $125

PHILCO

131, c.1946
BAKELITE, 4 TUBES, 1 BAND
$40

PHILCO

132, c.1946
WOOD, 5 TUBES, 1 BAND
$60

PHILCO

144B, c.1935
WOOD, 6 TUBES, 4 BANDS
$40

PHILCO

150, c.1948

WOOD, 5 TUBES, 1 BAND

$35

PHILCO

204, c.1947

WOOD, 5 TUBES, 1 BAND

$75

PHILCO

205, c.1947

WOOD, 5 TUBES, 1 BAND

$75

PHILCO

214, c.1947

WOOD, 5 TUBES, 1 BAND

$85

PHILCO

335T, c.1942

WOOD, 7 TUBES, 1 BAND

$85

PHILCO

350, c.1946

WOOD, 6 TUBES, 1 BAND

$65

PHILCO

442, c.1946

WOOD, 7 TUBES, AM/FM

$60

PHILCO

460, c.1948

BAKELITE, 6 TUBES, 1 BAND

$80

PHILCO

461, c.1948

WOOD, 6 TUBES, 1 BAND

$40

PHILCO

464, c.1948

BAKELITE, 6 TUBES, 1 BAND

$45

PHILCO

472, c.1948

WOOD, 8 TUBES, AM/FM

$45

PHILCO

475, c.1948

WOOD, 8 TUBES, AM/FM

$90

PHILCO

482, c.1948
WOOD, 9 TUBES, AM/FM/SW
$45

PHILCO

500, c.1949
BAKELITE, 5 TUBES, 1 BAND
$35

PHILCO

501, RADIO-PHONO, c.1940
WOOD, 5 TUBES, 1 BAND
$65

PHILCO

502, RADIO-PHONO, c.1940
WOOD, 5 TUBES, 1 BAND
$75

PHILCO

503, RADIO-PHONO, c.1940
WOOD, 6 TUBES, 1 BAND
$50

PHILCO

504, c.1941
BAKLELITE, 5 TUBES, 1 BAND
$40

PHILCO

527, c.1950
BAKLELITE, 5 TUBES, 1 BAND
$125

PHILCO

579 (BROWN), 590(BLACK),
591(RED), c.1955
PLASTIC, 4 TUBES, 1 BAND
BROWN $25, BLACK $25,
RED $55

PHILCO

602P, RADIO-PHONO, c.1941
WOOD, 5 TUBES, 1 BAND
$50

PHILCO

603, c.1949
5 TUBES, 1 BAND
$65

PHILCO

603P, RADIO-PHONO, c.1941
WOOD, 6 TUBES, 2 BANDS
$55

PHILCO

605, c.1949
6 TUBES, 1 BAND
$60

PHILCO

610T, 611T, c.1937
WOOD, 5 TUBES, 3 BANDS
$165

PHILCO

620P "BEAM-O-LIGHT", c.1942
WOOD, PHONO ONLY, 6 TUBES
$125

PHILCO

635B, c.1936
WOOD, 6 TUBES, 2 BANDS
$150

PHILCO

640, c.1952
PLASTIC, 5 TUBES, 1 BAND
RED $45

PHILCO

641, c.1952
PLASTIC, 6 TUBES, 1 BAND
GREEN, RED, BLUE $75
TAN $50

PHILCO

643, c.1952
PLASTIC, 6 TUBES, 1 BAND
TAN $35

PHILCO

645B, c.1936
WOOD, 7 TUBES, 2 BANDS
$150

PHILCO

681, c.1958
PLASTIC, 5 TUBES, 1 BAND
$45

PHILCO

706, RADIO-PHONO, c.1949
WOOD, 5 TUBES, 1 BAND
$65 *

*RECORD PLAYER MECHANISM RARELY FUNCTIONAL

PHILCO

717(BROWN), 719(IVORY), c.1955
PLASTIC, 4 TUBES, ONE BAND
BROWN, IVORY $40

PHILCO

726, c.1955
PLASTIC, 4 TUBES, 1 BAND
BLACK $40

PHILCO

740, c.1956
PLASTIC, 5 TUBES, 1 BAND
$45

PHILCO
742, c.1956
PLASTIC, 5 TUBES, 1 BAND
IVORY $35, PINK, $50

PHILCO
749, c.1959
PLASTIC, 5 TUBES, 1 BAND
IVORY $65

PHILCO
750, c.1958
PLASTIC, 5 TUBES, 1 BAND
IVORY, TAN $40, PINK $60

PHILCO
751, c.1959
PLASTIC, 5 TUBES, 1 BAND
IVORY $65, PINK $110, BLUE $150

PHILCO
753, c.1959
PLASTIC, 5 TUBES, 1 BAND
IVORY, BLACK $35, PINK $50

PHILCO
754, c.1958
PLASTIC, 5 TUBES, 1 BAND
IVORY, GREY $75

PHILCO
755, c.1959
PLASTIC, 5 TUBES, 1 BAND
IVORY $55, BLUE $90

PHILCO
758, c.1958
PLASTIC, 5 TUBES, 1 BAND
TAN $75, PINK $125

PHILCO
761, c.1959
PLASTIC, 5 TUBES, 1 BAND
BROWN $45, PINK $80

PHILCO
769, c.1960
PLASTIC, 5 TUBES, 1 BAND
BLUE $60

PHILCO
772, c.1960
PLASTIC, 5 TUBES, 1 BAND
TAN $40, PINK $60, BLUE $75

PHILCO
777, c.1961
PLASTIC, 5 TUBES, 1 BAND
IVORY $45

PHILCO

778, c.1961

PLASTIC, 5 TUBES, 1 BAND

GREY $40, BROWN $35

PHILCO

780, c.1961

PLASTIC, 5 TUBES, 1 BAND

BROWN $35, IVORY $40

PHILCO

782, c.1961

PLASTIC, 5 TUBES, 1 BAND

BROWN $35, AQUA $65

PHILCO

783, c.1961

PLASTIC, 5 TUBES, 1 BAND

TAN $45, PINK $75

PHILCO

784, c.1961

PLASTIC, 5 TUBES, 1 BAND

BROWN, IVORY, BLACK $40

AQUA $65

PHILCO

785, "PREDICTA", c.1961

PLASTIC, 5 TUBES, 1 BAND

BLACK, IVORY $150

PHILCO

788T, c.1941

WOOD

$55

PHILCO

790, c.1962

PLASTIC, 5 TUBES, 1 BAND

IVORY $40

PHILCO

792, c.1962

PLASTIC, 5 TUBES, 1 BAND

BLACK, BROWN, IVORY $45

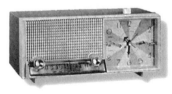

PHILCO

795, c.1962

PLASTIC, 5 TUBES, 1 BAND

BROWN $40, BLUE $60

PHILCO

796, c.1962

PLASTIC, 5 TUBES, 1 BAND

GREY, IVORY $40

PHILCO

797, "PREDICTA", c.1962

PLASTIC, 5 TUBES, 1 BAND

BLACK, IVORY $150

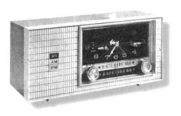

PHILCO

799, c.1962
PLASTIC, 7 TUBES, AM/FM
TAN, BROWN $40

PHILCO

806, c.1946
BAKELITE, 5 TUBES, 1 BAND
$50

PHILCO

808, c.1956
BAKELITE, 5 TUBES, 1 BAND
$45

PHILCO

809, c.1957
BAKELITE, 5 TUBES, 1 BAND
IVORY, BROWN $35

PHILCO

812, c.1956
PLASTIC, 5 TUBES, 1 BAND
YELLOW $60, ORANGE $70

PHILCO

813, c.1957
PLASTIC, 5 TUBES, 1 BAND
BROWN $35, PINK $60
YELLOW $60

PHILCO

814, c.1956
PLASTIC, 5 TUBES, 1 BAND
IVORY $40, BROWN $35,
GREEN $65, RED $70

PHILCO

814, c.1956
PLASTIC, 5 TUBES, 1 BAND
IVORY $40, GREY $45
RED $70

PHILCO

816, c.1956
PLASTIC, 5 TUBES, 1 BAND
IVORY $55, TAN $65

PHILCO

816, 'TROPIC', c.1946
WOOD, 5 TUBES, 4 BANDS
$110

PHILCO

817, c.1957
PLASTIC, 5 TUBES, 1 BAND
BROWN $45, TAN $65

PHILCO

820, c.1959
PLASTIC, 5 TUBES, 1 BAND
$50

PHILCO

822, C.1959

PLASTIC, 5 TUBES, 1 BAND
IVORY $55, YELLOW $85

PHILCO

824, C.1959

PLASTIC, 5 TUBES, 1 BAND
IVORY $45, PINK $65

PHILCO

826, C.1959

PLASTIC, 5 TUBES, 1 BAND
IVORY $45, ORANGE $75
BLUE $75

PHILCO

828, C.1959

PLASTIC, 5 TUBES, 1 BAND
IVORY $45, BROWN $40

PHILCO

829, C.1959

PLASTIC, 5 TUBES, 1 BAND
IVORY $55, YELLOW $85

PHILCO

847, C.1961

PLASTIC, 5 TUBES, 1 BAND
BLACK $45

PHILCO

849, C.1961

PLASTIC, 5 TUBES, 1 BAND
TAN $45, BROWN $35

PHILCO

850, C.1961

PLASTIC, 5 TUBES, 1 BAND
BROWN $40, IVORY $45,
GREY $50, BLUE $75

PHILCO

851, C.1961

PLASTIC, 5 TUBES, 1 BAND
BROWN $40, BLACK $55
YELLOW $85

PHILCO

852, C.1961

PLASTIC, 5 TUBES, 1 BAND
BROWN $40, PINK $65

PHILCO

853, C.1961

PLASTIC, 5 TUBES, 1 BAND
TAN $40, AQUA $65

PHILCO

855, C.1961

PLASTIC, 5 TUBES, 1 BAND
BROWN $40, AQUA $65

PHILCO
856, c.1961
PLASTIC, 5 TUBES, 1 BAND
BROWN $40, BLACK $45
AQUA $65

PHILCO
860, c.1962
PLASTIC, 5 TUBES, 1 BAND
BLACK $50

PHILCO
865, c.1962
PLASTIC, 5 TUBES, 1 BAND
IVORY $40, BLACK $45

PHILCO
866, c.1961
PLASTIC, 5 TUBES, 1 BAND
IVORY $40

PHILCO
901, 'SECRETARY', c.1949
BAKELITE, 5 TUBES, 1 BAND
BROWN, IVORY $225

PHILCO
904, c.1949
BAKELITE, 6 TUBES, 2 BANDS
$45

PHILCO
905, c.1949
BAKELITE, 6 TUBES, AM/FM
$40

PHILCO
906, c.1949
BAKELITE, 8 TUBES, AM/FM
$60

PHILCO
912, c.1961
PLASTIC, FM ONLY
TAN $40, IVORY $35
BLACK $40, AQUA $60

PHILCO
914, c.1961
PLASTIC, AM/FM
TAN $40, IVORY $35
BLACK $40, AQUA $60

PHILCO
925, c.1950
BAKELITE, 6 TUBES, AM/FM
$40

PHILCO
926,927, c.1961
PLASTIC, 7 TUBES, AM/FM
BROWN $35, BLUE $65

PHILCO

928, c.1962

WOOD, 7 TUBES, AM/FM
$40

PHILCO

929, c.1962

WOOD, 7 TUBES, AM/FM
$90

PHILCO

940, c.1952

BAKELITE, 6 TUBES, 1 BAND
$110

PHILCO

941, c.1952

BAKELITE, 6 TUBES, 1 BAND
$85

PHILCO

942, c.1952

BAKELITE, 6 TUBES, 1 BAND
$85

PHILCO

944, c.1952

BAKELITE, 7 TUBES, AM/FM
$50

PHILCO

963, c.1959

PLASTIC, 6 TUBES, 1 BAND
IVORY, BROWN $45

PHILCO

974, c.1958

HENRY DREYFUSS DES.(ATTRIB.)
PLASTIC, 7 TUBES, AM/FM
$50

PHILCO

976, c.1957

WOOD, 7 TUBES, AM/FM
$30

PHILCO

978, c.1959

WOOD, 8 TUBES, AM/FM
$50

PHILCO

995, c.1959

PLASTIC, 6 TUBES, FM ONLY
$40

PHILCO

996, c.1959

PLASTIC, 7 TUBES, AM/FM
$40

PHILCO
1001, RADIO-PHONO, C.1942
WOOD, 5 TUBES, 1 BAND
$40

PHILCO
1002, RADIO-PHONO, C.1942
WOOD, 6 TUBES, 1 BAND
$40

PHILCO
1003, RADIO-PHONO, C.1942
WOOD, 7 TUBES, 2 BANDS
$55

PHILCO
1201, RADIO-PHONO, C.1946
WOOD, 5 TUBES, 1 BAND
$125

PHILCO
1202, RADIO-PHONO, C.1946
WOOD, 6 TUBES, 1 BAND
$100

PHILCO
1203, RADIO-PHONO, C.1946
WOOD, 6 TUBES, 1 BAND
$75

PHILCO
1204, RADIO-PHONO, C.1946
WOOD, 6 TUBES, AM/FM
$85

PHILCO
1253, RADIO-PHONO, C.1948
WOOD, 5 TUBES, 1 BAND
$75

PHILCO
1256, RADIO-PHONO, C.1948
WOOD, 6 TUBES, 1 BAND
$50

PHILCO
1340, RADIO-PHONO, C.1952
BAKELITE, 5 TUBES, 1 BAND
$45

PHILCO
1401, RADIO-PHONO, C.1949
'BOOMERANG'
BAKELITE, 5 TUBES, 1 BAND
$225

PHILCO
1405, RADIO-PHONO, C.1949
WOOD, 5 TUBES, 1 BAND
$40

PHILCO

1408, RADIO-PHONO, C.1957
WOOD, 5 TUBES, 1 BAND
$40

PHILCO

1409, RADIO-PHONO, C.1958
WOOD, 5 TUBES, 1 BAND
$40

PHILCO

2465, 'AUTO CONTROL', C.1945
WOOD, CLOCK/TIMER
$75

PHILCO

2670B, C.1938
WOOD, 11 TUBES, 5 BANDS
$245

PHILCO

B649, C.1953
PLASTIC, 4 TUBES, 1 BAND, DC
$70

PHILCO

'COLONIAL CLOCK', C.1932
WOOD, 5 TUBES, 1 BAND
$350

PHILCO

E812-124, C.1957
PLASTIC, 5 TUBES, 1 BAND
$45

PHILCO

EXTENSION SPEAKER, C.1932
WOOD, SPEAKER ONLY
$150

PHILCO

M15, PHONO REMOTE, C.1950
BAKELITE, PHONO ONLY
$75

PHILCO

PT-2, C.1941
BAKELITE, 5 TUBES, 1 BAND
$50

PHILCO

PT-6, C.1941
WOOD, 5 TUBES, 1 BAND
$85

PHILCO

PT-30, C.1941
BAKELITE, 5 TUBES, 1 BAND
$60

PHILCO

PT-42, c.1941
WOOD, 5 TUBES, 1 BAND
$65

PHILCO

PT-44, c.1941
WOOD, 5 TUBES, 1 BAND
$125

PHILCO

PT-61, 'PAGODA', c.1941
WOOD, 5 TUBES, 1 BAND
$375

PHILCO

PT-202, c.1946
WOOD, 5 TUBES, 1 BAND
$110

PHILCO

PT-540, c.1952
BAKELITE, 5 TUBES, 1 BAND
BROWN, IVORY $45

PHILCO

PT-542, c.1952
BAKELITE, 5 TUBES, 1 BAND
BROWN, IVORY $65

PHILCO

PT-544, c.1952
BAKELITE, 5 TUBES, 1 BAND
BROWN, IVORY $90

PHILCO

PT-548, c.1952
BAKELITE, 5 TUBES, 1 BAND
BROWN, IVORY $125

PHILCO

RP1, PHONO REMOTE, c.1942
WOOD, PHONO ONLY
$50

PHILCO

T-600, c.1959
PLASTIC, TRANSISTORIZED, 1 BAND
$75

PHILCO

T-900, c.1959
PLASTIC, TRANSISTORIZED, 1 BAND
$65

PHILCO
CANADA

29T, c.1940
WOOD, 5 TUBES, 1 BAND
$70

PHILCO
CANADA
57, C.1949
BAKELITE, 5 TUBES, 1 BAND
$50

PHILCO
CANADA
58, 'BOOMERANG', C.1949
WOOD, 5 TUBES, 1 BAND
$125

PHILCO
CANADA
61, C.1949
BAKELITE, 5 TUBES, 1 BAND
$50

PHILCO
CANADA
65, C.1949
WOOD, 6 TUBES, 1 BAND
$45

PHILCO
CANADA
66, C.1949
WOOD, 6 TUBES, 1 BAND
$60

PHILCO
CANADA
74, C.1949
BAKELITE, 6 TUBES, 1 BAND
$75

PHILCO
CANADA
76, C.1949
BAKELITE, 6 TUBES, 1 BAND
$75

PHILCO
CANADA
81, C.1950
BAKELITE, 6 TUBES, 1 BAND
$40
PAINTED: IVORY; BROWN; GREY; RED;
BLUE; GREEN

PHILCO
CANADA
82, C.1950
BAKELITE, 5 TUBES, 1 BAND
BROWN, IVORY $40

PHILCO
CANADA
87, C.1950
BAKELITE, 5 TUBES, 1 BAND
BROWN, IVORY, RED $40

PHILCO
CANADA
97, C.1950
BAKELITE, 5 TUBES, 1 BAND
BROWN, GREEN, RED $90

PHILCO
CANADA
99, C.1950
BAKELITE, 5 TUBES, 1 BAND
BROWN, IVORY $75

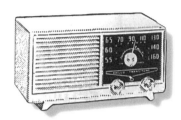

PHILCO
CANADA
162, c.1953
BAKELITE, 5 TUBES, 1 BAND
$40

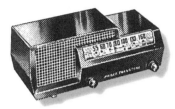

PHILCO
CANADA
163, 'SPLIT LEVEL', c.1953
BAKELITE, 5 TUBES, 1 BAND
$125

PHILCO
CANADA
164, c.1953
BAKELITE, 5 TUBES, 1 BAND
$45

PHILCO
CANADA
166, c.1952
BAKELITE, 5 TUBES, 1 BAND
$125

PHILCO
CANADA
170, c.1954
BAKELITE, 4 TUBES, 1 BAND
$50

PHILCO
CANADA
174, c.1954
BAKELITE, 5 TUBES, 1 BAND
$35

PHILCO
CANADA
182, c.1955
BAKELITE, 5 TUBES, 1 BAND
$60

PHILCO
CANADA
217, c.1949
BAKELITE, 4 TUBES, 1 BAND, DC
$30

PHILCO
CANADA
218, c.1949
WOOD, 5 TUBES, 1 BAND, DC
$40

PHILCO
CANADA
219, c.1949
WOOD, 5 TUBES, 1 BAND, DC
$40

PHILCO
CANADA
222, c.1949
WOOD, 5 TUBES, 1 BAND, DC
$30

PHILCO
CANADA
225, c.1950
BAKELITE, 5 TUBES, 1 BAND, DC
IVORY, BROWN, RED, GREEN $40

PHILCO
CANADA
315, C.1955
PLASTIC, 6 TUBES, 1 BAND
$45

PHILCO
CANADA
409, C.1949
PLASTIC, 6 TUBES, 2 BANDS
$45

PHILCO
CANADA
410, 411, C.1949
PLASTIC, 5 TUBES, 1 BAND
IVORY $40, TAN $45
RED $90, GREEN $80

PHILCO
CANADA
414, C.1950
PLASTIC, 4 TUBES, 1 BAND
BLUE $90, TAN $65
RED $85, GREEN $85

PHILCO
CANADA
452, C.1954
PLASTIC, 4 TUBES, 1 BAND
IVORY $70, BLUE $100
RED $90, GREEN $90

PHILCO
CANADA
466, C.1955
PLASTIC, 5 TUBES, 1 BAND
$55

PHILCO
CANADA
708, RADIO-PHONO, C.1949
WOOD, 6 TUBES, 1 BAND
$75

PHILIPS
CANADA
378, C.1958
WOOD, 9 TUBES, 3 BANDS + FM
$65

PHILIPS
CANADA
473, C.1958
WOOD, 10 TUBES, 3 BANDS + FM
$85

PHILIPS
CANADA
725, C.1954
WOOD, 6 TUBES, 2 BANDS
$65

PHILIPS
CANADA
735, C.1952
WOOD, 4 TUBES, 1 BAND
$50

PHILIPS
CANADA
825, C.1952
WOOD, 6 TUBES, 1 BAND
$60

PHILIPS
C A N A D A
905, c.1949
BAKELITE, 5 TUBES, 1 BAND
$60

PHILIPS
C A N A D A
922, c.1952
BAKELITE, 5 TUBES, 1 BAND
$65

PHILIPS
C A N A D A
924, c.1953
WOOD, 5 TUBES, 1 BAND
$65

PHILIPS
C A N A D A
B2-C1OU, c.1958
PLASTIC, 5 TUBES, 1 BAND
ORANGE, AQUA $150

PHILIPS
C A N A D A
CM23L, c.1952
BAKELITE, 5 TUBES, 1 BAND
$45

PHILIPS
C A N A D A
CM33A, c.1958
WOOD, 11 TUBES, 5 BANDS
$225

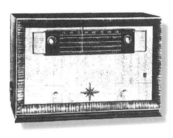

PHILIPS
C A N A D A
PH103, c.1954
WOOD, 8 TUBES, 4 BANDS
$75

PHILMORE
'ONE TUBE', c.1932
CARDBOARD, 1 TUBE, 1 BAND
$125

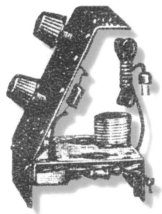

PHILMORE
ONE TUBE KIT, c.1942
ONE TUBE, ONE BAND
$50

PHILMORE
TWO TUBE KIT, c.1940
ONE TUBE, ONE BAND
$65

PHILMORE
TWO TUBE KIT, c.1942
ONE TUBE, ONE BAND
$50

PHILMORE
'TWO TUBE', c.1932
CARDBOARD, 2 TUBES, 1 BAND
$125

PHILMORE

'BLACKBIRD', C.1932

CRYSTAL

$250

PHILMORE

'DWARF', C.1942

ONE TUBE, ONE BAND

$75

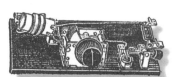

PHILMORE

'GIANT', C.1942

CRYSTAL

$75

PHILMORE

'LITTLE GIANT', C.1942

CRYSTAL

$85

PHILMORE

'LITTLE WONDER', C.1932

CRYSTAL

$125

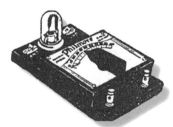

PHILMORE

'MONARCH', C.1942

CRYSTAL

$75

PHILMORE

'SELECTIVE', C.1932

CRYSTAL

$145

PHILMORE

'SUPERTONE', C.1932

CRYSTAL

$85

PHILMORE

'TINY TONE', C.1942

VARIABLE CRYSTAL

$100

PHILMORE

'TOP NOTCH', C.1942

CRYSTAL

$75

PHILMORE

'ULTRATONE', C.1932

CRYSTAL

$110

"Phonola"

'CHATTER BOX', C.1940

WOOD, 5 TUBES, 1 BAND

$50

"Phonola"

'MANTLE', RADIO-PHONO, C.1940
WOOD, 5 TUBES, 1 BAND
$55

PILOT

20, C.1937
WOOD, 7 TUBES, ALL-WAVE
$350

PILOT

63, C.1935
WOOD, 6 TUBES, ALL-WAVE
$325

PILOT

114, C.1935
WOOD
$300

PILOT

B1, C.1941
BAKELITE/PLASKON
BROWN $100, IVORY $150

PILOT

K115, C.1931
'SUPERWASP'
$350

PILOT

'PILOT 9', C.1932
WOOD, 6 TUBES, ALL-WAVE
$275

PILOT

SW CONVERTER, C.1932
WOOD, 4 TUBES, SHORT WAVE
$150

PILOT

T-301, C.1942
WOOD, 8 TUBES, AM/FM
$50

PILOT

T-47, C.1941
WOOD
$150

PLYMOUTH

'MIGHTY MIDGET', C.1932
WOOD, 5 TUBES, 1 BAND
$225

PLYMOUTH

'SILENTUNE', RADIO-PHONO
C.1933
WOOD, 5 TUBES, 1 BAND
$375

PLYMOUTH

'Silentinting', c.1933
WOOD, 5 TUBES, 1 BAND
$225

POLICE ALARM

PR-31, c.1951
BAKELITE, 5 TUBES, POLICE BAND
$40

PORTO

SR-600, 'Smokerette'
c.1947
BAKELITE, 5 TUBES, 1 BAND
$125

POWERTONE

'One Tube', c.1933
WOOD, 1 TUBE, 1 BAND
$50

POWERTONE

8T-1964, 'Royal', c.1933
WOOD, 4 TUBES, 1 BAND
$85

POWERTONE

5104, c.1933
WOOD, 5 TUBES, 1 BAND
$145

POWERTONE

C5, 'Rochambeau', c.1933
WOOD, 5 TUBES, 1 BAND
$115

POWERTONE

CR4, 'Cadet', c.1933
WOOD, 4 TUBES, 1 BAND
CRYSTAL FINISH $250

POWERTONE

EUS4, 'Reliable', c.1933
WOOD, 4 TUBES, 1 BAND
$75

POWERTONE

JR4, 'Junior', c.1933
WOOD, 4 TUBES, 1 BAND
$85

POWERTONE

LA5S, 'Rochambeau', c.1934
WOOD, 5 TUBES, 1 BAND
$135

POWERTONE

MBS1W, 'Master', c.1933
WOOD, 1 TUBE, 1 BAND
$75

POWERTONE

'PAL', C.1933

WOOD, REMOTE SPEAKER
$50

POWERTONE

PB6(DC), PR5W(AC), C.1933

WOOD, 5 TUBES, 1 BAND
DC $100, AC $200

POWERTONE

PMR6, 'PARAMOUNT', C.1933

WOOD, 6 TUBES, 1 BAND
$110

POWERTONE

PB6M(DC), PR5WM(AC)
C.1933

WOOD, 5 TUBES, 1 BAND
DC $40, AC $85

POWERTONE

PRF4, R6, C.1933

WOOD, 4/6 TUBES, 1 BAND
$200

POWERTONE

PTR1960, C.1933
'TREASURE CHEST'

WOOD, 5 TUBES, 1 BAND
$250

POWERTONE

PTU1964, C.1933

WOOD, 5 TUBES, 1 BAND
$75

POWERTONE

PUM5' LA MODERNE', C.1934

WOOD, 5 TUBES, 1 BAND
$175

POWERTONE

RMR3018, C.1933

WOOD, 4 TUBES, 1 BAND
$600

POWERTONE

SW ADAPTER, C.1933

WOOD, SHORT WAVE
$65

POWERTONE

SW5708, C.1933

WOOD, 2 TUBES, SHORT WAVE
$75

POWERTONE

TR1961/2, C.1933

WOOD, 4/5 TUBES, 1 BAND
$200

Puretone

'Adjustable Crystal', c.1935

CRYSTAL

$65

PYE

39HC, c.1951

WOOD, 6 TUBES, 8 BANDS

$125

Radionic

'Chancellor', c.1947

WOOD, 6 TUBES, 1 BAND

$35

RADOLEK

10970, c.1935

WOOD, 8 TUBES, 2 BANDS

$300

RADOLEK

945, SW Converter, c.1933

WOOD, 2 TUBES, SHORTWAVE

$75

RADOLEK

950, c.1933

WOOD, 4 TUBES, 1 BAND

$125

RADOLEK

951, 'Super Duola', c.1933

WOOD, 5 TUBES, 2 BANDS

$85

RADOLEK

953, 'Companion', c.1933

WOOD, 4 TUBES, 1 BAND

$225

RADOLEK

954, c.1933

WOOD, 4 TUBES, 1 BAND

$225

RADOLEK

956, c.1933

WOOD, 6 TUBES, 2 BANDS

$135

RADOLEK

957, 'Duola', c.1933

WOOD, REMOTE SPEAKER

$75

RADOLEK

958, 'Deluxe Super 6', c.1933

WOOD, 6 TUBES, 4 BANDS

$150

RADOLEK

960, 'FIRESIDE', C.1933
WOOD, 5 TUBES, 2 BANDS, DC
$100

RADOLEK

963, C.1933
WOOD, 8 TUBES, 2 BANDS, DC
$100

RADOLEK

967, C.1933
WOOD, 8 TUBES, 1 BANDS, DC
$100

RADOLEK

989, 'DELUXE', C.1933
WOOD, 6 TUBES, 2 BANDS
$225

RADOLEK

6785, C.1937
WOOD, 6 TUBES, 2 BANDS
$85

RADOLEK

10970, C.1935
WOOD, 8 TUBES, 2 BANDS
$325

RADOLEK

16700, C.1937
WOOD, 4 TUBES, 2 BANDS
$70

RADOLEK

16701, C.1937
WOOD, 5 TUBES, 2 BANDS
$110

RADOLEK

16702, C.1937
WOOD, 5 TUBES, 2 BANDS, DC
$110

RADOLEK

16703, C.1937
WOOD, 7 TUBES, 3 BANDS, DC
$90

RADOLEK

16705, C.1937
WOOD, 7 TUBES, 2 BANDS
$225

RADOLEK

16706, 'ZEPHYR', C.1937
WOOD, 6 TUBES, 3 BANDS
$650

RADOLEK

16721/4, c.1937
WOOD, 7/10 TUBES, 3 BANDS
7 TUBES $150
10 TUBES $250

RADOLEK

16727, c.1937
WOOD, 5 TUBES, 2 BANDS
$350

RADOLEK

16740, c.1937
WOOD, 6 TUBES, 2 BANDS
$125

RADOLEK

16745, c.1937
WOOD, 7 TUBES, 4 BANDS, DC
$150

RADOLEK

16747, c.1937
WOOD, 5 TUBES, 3 BANDS
$95

RADOLEK

16748, c.1937
WOOD, 5 TUBES, 2 BANDS
$125

RADOLEK

16755, c.1937
WOOD, 6 TUBES, 3 BANDS
$200

RADOLEK

16758, c.1937
WOOD, 8 TUBES, 4 BANDS
$375

RADOLEK

16771, c.1937
WOOD, 8 TUBES, 3 BANDS
$275

RADOLEK

16774, c.1937
WOOD, 6 TUBES, 2 BANDS, DC
$65

RADOLEK

16776, c.1937
WOOD, 11 TUBES, 4 BANDS
$425

RADOLEK

16779, c.1937
WOOD, 8 TUBES, 3 BANDS, DC
$110

RADOLEK

16791, C.1937
WOOD, 7 TUBES, 3 BANDS
$325

RADOLEK

B17574, C.1939
BAKELITE, 6 TUBES, 1 BAND
$125

RADOLEK

B17658, C.1939
BAKELITE, 5 TUBES, 1 BAND
$350

RADOLEK

B17660, C.1939
BAKELITE, 5 TUBES, 1 BAND
$225

RADOLEK

B17661(DC), B17678 (AC) C.1939
BAKELITE, 5 TUBES, 1 BAND
$175

RADOLEK

B17663, C.1939
BAKELITE, 5 TUBES, 1 BAND, DC
$50

RADOLEK

B17665, C.1939
BAKELITE, 5 TUBES, 1 BAND, DC
$60

RADOLEK

B17667, C.1939
BAKELITE, 6 TUBES, 2 BANDS
$115

RADOLEK

B17668, C.1939
BAKELITE, 7 TUBES, 2 BAND
$75

RADOLEK

B17674, C.1939
BAKELITE, 2 TUBES, 1 BAND
$175

RADOLEK

B17676, C.1939
BAKELITE, 6 TUBES, 1 BAND
$200

RADOLEK

B17683, C.1939
METAL, 5 TUBES, 1 BAND
$175

RCA Victor
c.1934
WOOD, 4 TUBES, 1 BAND
$75

RCA Victor
1AX(BROWN), 1AX2(IVORY), c.1940
BAKELITE, 5 TUBES, 1 BAND
$60

RCA Victor
1X, c.1940
BAKELITE, 5 TUBES, 1 BAND
$60

RCA Victor
3RF91, 'WOODARD', c.1952
BAKELITE, 5 TUBES, 1 BAND
$35

RCA Victor
4C531, 'REVEILLE', c.1952
BAKELITE, 5 TUBES, 1 BAND
$35

RCA Victor
4C671, 'PROMPTER', c.1952
BAKELITE, 5 TUBES, 1 BAND
$35

RCA Victor
4V541, 'SLUMBERETTE', c.1952
BAKELITE, 5 TUBES, 1 BAND
$35

RCA Victor
4X555, 'CREIGHTON', c.1957
PLASTIC, 5 TUBES, 1 BAND
BROWN $65, BLACK $115
GREEN $135, RED $135

RCA Victor
5Q1, c.1938
WOOD, 5 TUBES, 3 BANDS
$95

RCA Victor
5Q2, c.1938
WOOD, 5 TUBES, 3 BANDS
$115

RCA Victor
5Q4, c.1938
WOOD, 5 TUBES, 4 BANDS
$85

RCA Victor
5Q5, 5Q55, 5Q56, c.1938
JOHN VASSOS DES.(ATTRIB.)
WOOD, 5 TUBES, 3 BANDS
$110

RCA Victor
5Q6, C.1939
JOHN VASSOS DES.(ATTRIB.)
WOOD, 5 TUBES, 3 BANDS
$110

RCA Victor
5Q8, C.1939
JOHN VASSOS DES.(ATTRIB.)
WOOD, 5 TUBES, 3 BANDS
$110

RCA Victor
5T, C.1937
WOOD, 5 TUBES, 2 BANDS
$125

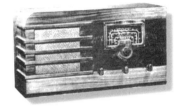

RCA Victor
5T4, C.1937
WOOD, 5 TUBES, 3 BANDS
$115

RCA Victor
5T5, C.1937
WOOD, 5 TUBES, 2 BANDS
$135

RCA Victor
5T6, 5T7(IVORY), C.1937
JOHN VASSOS DES.(ATTRIB.)
WOOD, 5 TUBES, 2 BANDS
$175

RCA Victor
5T8, C.1937
JOHN VASSOS DES.(ATTRIB.)
WOOD, 5 TUBES, 2 BANDS
$225

RCA Victor
5U, RADIO-PHONO, C.1936
WOOD, 5 TUBES, 2 BANDS
$150

RCA Victor
6Q1, C.1939
JOHN VASSOS DES.(ATTRIB.)
BAKELITE, 6 TUBES, 3 BANDS
$110

RCA Victor
6Q4, C.1939
JOHN VASSOS DES.(ATTRIB.)
BAKELITE, 6 TUBES, 4 BANDS
$110

RCA Victor
6Q7, C.1939
WOOD, 6 TUBES, 3 BANDS
$70

RCA Victor
6QU, RADIO-PHONO, C.1939
WOOD, 5 TUBES, 3 BANDS
$85

RCA Victor
6T, c.1937
WOOD, 6 TUBES, 2 BANDS
$135

RCA Victor
6T5, c.1937
WOOD, 6 TUBES, 2 BANDS
$145

RCA Victor
6T10, c.1937
JOHN VASSOS DES.
WOOD & CHROME, 6 TUBES, 3 BANDS
$3,500

RCA Victor
6X, c.1942
WOOD, 6 TUBES, 1 BAND
$45

RCA Victor
6X13, c.1940
WOOD, 6 TUBES, 2 BANDS
$50

RCA Victor
7Q4, c.1939
WOOD, 7 TUBES, 4 BANDS
$60

RCA Victor
7T, 7X, c.1936
WOOD, 7 TUBES, 3 BANDS
$175

RCA Victor
7T1, c.1936
WOOD, 7 TUBES, 3 BANDS
$200

RCA Victor
7X1, c.1936
WOOD, 7 TUBES, 2 BANDS
$175

RCA Victor
8BT, 8BT6, c.1936
WOOD, 8 TUBES, 3 BANDS, DC
$100

RCA Victor
8BX6, c.1947
METAL, 6 TUBES, 1 BAND
$45

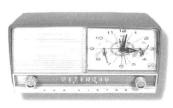

RCA Victor
8C-7LE, c.1959
PLASTIC, 5 TUBES, 1 BAND
$45

RCA Victor

8F43, c.1948

WOOD, 4 TUBES, 1 BAND, DC

$25

RCA Victor

8Q1, c.1938

WOOD, 8 TUBES, 3 BANDS

$125

RCA Victor

8Q2, c.1938

WOOD, 8 TUBES, 3 BANDS

$85

RCA Victor

8Q4, c.1938

WOOD & CHROME, 8 TUBES, 4 BANDS

$275

RCA Victor

8QB, c.1938

WOOD, 8 TUBES, 3 BANDS

$250

RCA Victor

8QU5, RADIO-PHONO, c.1938

WOOD, 8 TUBES, 3 BANDS

$150

RCA Victor

8R72, c.1947

BAKELITE, 7 TUBES, AM/FM

$45

RCA Victor

8T2, c.1936

WOOD, 8 TUBES, 3 BANDS

$200

RCA Victor

8T10, c.1936

JOHN VASSOS DES.

WOOD & CHROME, 8 TUBES, 3 BANDS

$4,500

RCA Victor

8T11, c.1936

JOHN VASSOS DES.

WOOD & CHROME, 8 TUBES, 3 BANDS

$4,500

RCA Victor

8X53, 'LOG CABIN', c.1948

WOOD, 5 TUBES, 1 BAND

$55

RCA Victor

8X542, c.1948

BAKELITE, 5 TUBES, 1 BAND

$65

RCA Victor
8X545, C.1948
BAKELITE, 5 TUBES, 1 BAND
$65

RCA Victor
8X681, C.1948
HENRY DREYFUSS DES.
BAKELITE, 6 TUBES, 2 BANDS
$250

RCA Victor
8X71, C.1947
BAKELITE, 7 TUBES, AM/FM
$40

RCA Victor
9BX5, C.1950
PLASTIC, 5 TUBES, 1 BAND
$55

RCA Victor
9Q1, C.1939
WOOD, 9 TUBES, 7 BANDS
$60

RCA Victor
9Q4, C.1939
WOOD, 9 TUBES, 4 BANDS
$60

RCA Victor
9X, C.1938
WOOD, 4 TUBES, 1 BAND
$350

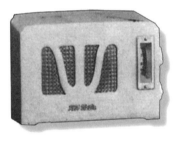

RCA Victor
9X11, C.1938
CATALIN, 4 TUBES, 1 BAND
GREEN, BLACK, IVORY, BROWN
$1,500+

RCA Victor
9X561, C.1950
BAKELITE, 5 TUBES, 1 BAND
$45

RCA Victor
9X641, C.1950
BAKELITE, 5 TUBES, 1 BAND
$35

RCA Victor
10Q1, C.1938
WOOD & CHROME, 10 TUBES, 3 BANDS
$375

RCA Victor
10T11, C.1938
JOHN VASSOS DES.
WOOD & CHROME, 10 TUBES, 5 BANDS
$5,000

RCA Victor
10X, 11X, c.1940
BAKELITE, 5 TUBES, 1 BAND
$35

RCA Victor
11Q4, c.1938
WOOD, 11 TUBES, 4 BANDS
$450

RCA Victor
12AX, 12AX2(IVORY), c.1940
BAKELITE, 5 TUBES, 1 BAND
$35

RCA Victor
12Q4, c.1938
WOOD, 12 TUBES, 8 BANDS
$575

RCA Victor
12X, c.1942
BAKELITE, 5 TUBES, 1 BANDS
$35

RCA Victor
14BT1, c.1940
BAKELITE, 4 TUBES, 1 BANDS, DC
$25

RCA Victor
14BT2, c.1940
WOOD, 4 TUBES, 1 BANDS, DC
$30

RCA Victor
14X, 14AX, c.1940
WOOD, 5 TUBES, 2 BANDS
$35

RCA Victor
15BP6, c.1940
WOOD, 5 TUBES, 1 BANDS
$40

RCA Victor
15BT, c.1940
WOOD, 5 TUBES, 2 BANDS, DC
$30

RCA Victor
16T2, c.1940
WOOD, 6 TUBES, 2 BANDS
$55

RCA Victor
16T3, c.1940
WOOD, 6 TUBES, 2 BANDS
$70

RCA Victor

16X3, c.1940

WOOD, 6 TUBES, 1 BAND

$40

RCA Victor

16X14, c.1940

WOOD, 6 TUBES, 2 BANDS

$45

RCA Victor

18T, c.1940

WOOD, 8 TUBES, 3 BANDS

$45

RCA Victor

24BT1, c.1940

BAKELITE, 4 TUBES, 1 BAND, DC

$25

RCA Victor

25BT2, c.1942

WOOD, 5 TUBES, 1 BAND, DC

$20

RCA Victor

26X1, c.1942

BAKELITE, 6 TUBES, 1 BAND, DC

$30

RCA Victor

26X3, c.1942

WOOD, 6 TUBES, 1 BAND, DC

$35

RCA Victor

26X4, c.1942

WOOD, 6 TUBES, 1 BAND, DC

$40

RCA Victor

28T, c.1942

WOOD, 8 TUBES, 1 BAND, DC

$40

RCA Victor

28X, c.1942

WOOD, 8 TUBES, 1 BAND, DC

$40

RCA Victor

40X-50, 'MODERN BLONDE', c.1939

WOOD, 5 TUBES, 1 BAND

$110

RCA Victor

40X-56, '1939 WORLD'S FAIR', c.1939

WOOD/REPWOOD, 5 TUBES, 1 BAND

$2,200

RCA Victor

45X2, c.1940

BAKELITE, 5 TUBES, 1 BAND

$75

RCA Victor

45X3, c.1940

WOOD, 5 TUBES, 1 BAND

$60

RCA Victor

45X4, c.1940

WOOD, 5 TUBES, 1 BAND

$65

RCA Victor

45X16, c.1940

WOOD, 5 TUBES, 1 BAND

$60

RCA Victor

45X17, c.1940

WOOD, 5 TUBES, 1 BAND

$70

RCA Victor

45X18, c.1940

WOOD, 5 TUBES, 1 BAND

$75

RCA Victor

55F, c.1945

WOOD, 5 TUBES, 1 BAND, DC

$25

RCA Victor

55X, c.1940

WOOD, 5 TUBES, 1 BAND

$65

RCA Victor

56X, c.1945

BAKELITE, 6 TUBES, 2 BANDS

$35

RCA Victor

56X10, c.1945

BAKELITE, 6 TUBES, 2 BANDS

$35

RCA Victor

56X2, c.1945

BAKELITE, 6 TUBES, 2 BANDS

$35

RCA Victor

56X3, c.1945

WOOD, 6 TUBES, 2 BANDS

$45

RCA Victor
56X5, C.1945
BAKELITE, 6 TUBES, 2 BANDS
$45

RCA Victor
61-8, C.1946
WOOD, 5 TUBES, 1 BAND
$45

RCA Victor
64F3, C.1946
WOOD, 4 TUBES, 1 BAND
$30

RCA Victor
65X8, C.1946
BAKELITE, 5 TUBES, 1 BAND
$35

RCA Victor
66X4, C.1946
WOOD, 5 TUBES, 1 BAND
$45

RCA Victor
68R3, C.1946
WOOD, 8 TUBES, AM/FM
$30

RCA Victor
68R4, C.1946
WOOD, 8 TUBES, AM/FM
$50

RCA Victor
85T5, C.1937
WOOD, 5 TUBES, 2 BANDS
$85

RCA Victor
86T4, C.1937
WOOD, 6 TUBES, 3 BANDS
$100

RCA Victor
86T44, C.1937
WOOD, 6 TUBES, 3 BANDS
$90

RCA Victor
86T6, C.1938
WOOD, 6 TUBES, 3 BANDS
$125

RCA Victor
86X, C.1937
WOOD, 6 TUBES, 2 BANDS
$70

RCA Victor

86X4, C.1937

WOOD, 6 TUBES, 3 BANDS

$75

RCA Victor

87X, C.1937

WOOD, 7 TUBES, 3 BANDS

$75

RCA Victor

91B, C.1934

METAL, 4 TUBES, 1 BAND

$145

RCA Victor

94BT1, C.1938

WOOD, 4 TUBES, 1 BAND, DC

$50

RCA Victor

94BT2, C.1938

WOOD, 4 TUBES, 2 BANDS, DC

$40

RCA Victor

94BT6, C.1938

WOOD, 4 TUBES, 1 BAND, DC

$40

RCA Victor

95T5, C.1938

WOOD, 5 TUBES, 1 BAND

$75

RCA Victor

95X11, C.1938

WOOD, 5 TUBES, 1 BAND

$90

RCA Victor

95X6, C.1938

WOOD, 5 TUBES, 1 BAND

$75

RCA Victor

96BT6, C.1938

WOOD, 6 TUBES, 2 BANDS, DC

$45

RCA Victor

96T, C.1938

WOOD, 6 TUBES, 1 BAND

$80

RCA Victor

96T1, C.1938

WOOD, 6 TUBES, 1 BAND

$140

RCA Victor

96T2, C.1938

WOOD, 6 TUBES, 2 BANDS

$110

RCA Victor

96T3, C.1938

WOOD, 6 TUBES, 3 BANDS

$90

RCA Victor

96T7, C.1938

WOOD & CHROME, 7 TUBES, 3 BANDS

$115

RCA Victor

97T, C.1938

WOOD, 7 TUBES, 3 BANDS

$120

RCA Victor

97T2, C.1938

WOOD, 7 TUBES, 3 BANDS

$80

RCA Victor

97X, C.1938

WOOD, 7 TUBES, 1 BAND

$85

RCA Victor

98X, C.1938

WOOD, 8 TUBES, 3 BANDS

$135

RCA Victor

99T, C.1938

WOOD, 9 TUBES, 3 BANDS

$160

RCA Victor

118, C.1934

WOOD, 5 TUBES, 2 BANDS

$140

RCA Victor

124, C.1934

WOOD/REPWOOD, 6 TUBES, 1 BAND

$450

RCA Victor

126B, C.1934

WOOD, 6 TUBES, 1 BAND, DC

$75

RCA Victor

127, C.1934

WOOD, 6 TUBES, 2 BANDS, DC

$150

RCA Victor

135B, C.1934

WOOD, 7 TUBES, 2 BANDS

$170

RCA Victor

511, C.1940

BAKELITE, 5 TUBES, 1 BAND

$50

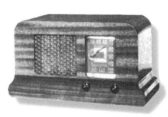

RCA Victor

512, C.1940

WOOD, 5 TUBES, 1 BAND

$60

RCA Victor

520, RADIOLA, C.1942

WOOD, 5 TUBES, 1 BAND

$45

RCA Victor

522, RADIOLA, C.1942

WOOD, 5 TUBES, 1 BAND

$40

RCA Victor

81OT4, C.1937

WOOD, 10 TUBES, 4 BANDS

$250

RCA Victor

812X, C.1937

WOOD, 12 TUBES, 4 BANDS

$300

RCA Victor

AVR5A, C.1934

9 TUBES, 5 BANDS

$150

RCA Victor

BP1O, C.1940

PLASTIC & METAL, 4 TUBES, 1 BAND

$40

RCA Victor

BT6-3, C.1936

WOOD, 6 TUBES, 1 BAND, DC

$60

RCA Victor

BT6-5, C.1936

WOOD, 6 TUBES, 2 BANDS, DC

$70

RCA Victor

BT7-8, C.1936

WOOD, 7 TUBES, 2 BANDS, DC

$75

RCA Victor
BT40, c.1939
WOOD, 4 TUBES, 1 BAND, DC
$40

RCA Victor
BT41, c.1940
WOOD, 4 TUBES, 1 BAND, DC
$20

RCA Victor
BT42, c.1939
WOOD, 4 TUBES, 1 BAND, DC
$25

RCA Victor
BT43, c.1940
WOOD, 4 TUBES, 1 BAND, DC
$20

RCA Victor
EXTENSION SPEAKER, c.1934
REPWOOD, SPEAKER ONLY
$150

RCA Victor
Q20, c.1940
BAKELITE, 5 TUBES, 2 BANDS
$50

RCA Victor
Q21, c.1940
WOOD, 5 TUBES, 2 BANDS
$50

RCA Victor
Q24, c.1940
BAKELITE, 6 TUBES, 5 BANDS
$65

RCA Victor
Q25, c.1940
WOOD, 6 TUBES, 5 BANDS
$60

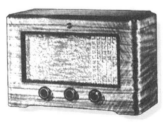

RCA Victor
Q26, c.1940
WOOD, 7 TUBES, 5 BANDS
$85

RCA Victor
Q33, c.1940
WOOD, 8 TUBES, 5 BANDS
$110

RCA Victor
Q44, c.1940
WOOD, 12 TUBES, 8 BANDS
$200

RCA Victor
QU3, c.1940
WOOD, 7 TUBES, 5 BANDS
$85

RCA Victor
R74, c.1933
WOOD, 10 TUBES, 2 BANDS
$325

RCA Victor
R91, PHONO, c.1938
WOOD, 4 TUBES, PHONO ONLY
$50

RCA Victor
RADIOLA 21, c.1932
WOOD
$125

RCA Victor
T4, c.1936
WOOD, 4 TUBES, 1 BAND
$110

RCA Victor
T6-7, c.1936
WOOD, 6 TUBES, 3 BANDS
$150

RCA Victor
T6-9, c.1936
WOOD, 6 TUBES, 2 BANDS
$110

RCA Victor
T6-11, c.1936
WOOD, 6 TUBES, 3 BANDS
$120

RCA Victor
T7-12, c.1936
WOOD, 7 TUBES, 3 BANDS
$160

RCA Victor
T8-18, c.1936
WOOD, 8 TUBES, 3 BANDS
$190

RCA Victor
T9-7,8, c.1936
WOOD, 9 TUBES, 3 BANDS
$240

RCA Victor
T9-10, c.1936
WOOD, 9 TUBES, 3 BANDS
$260

RCA Victor

T64, c.1939
WOOD, 6 TUBES, 3 BANDS
$45

RCA Victor

U8, RADIO-PHONO, c.1939
WOOD, 5 TUBES, 1 BAND
$65

RCA Victor

U9, RADIO-PHONO, c.1939
WOOD, 5 TUBES, 1 BAND
$50

RCA Victor

U10, RADIO-PHONO, c.1939
WOOD, 5 TUBES, 1 BAND
$55

RCA Victor

U12, RADIO-PHONO, c.1939
WOOD, 7 TUBES, 2 BANDS
$75

RCA Victor

U101, RADIO-PHONO, c.1937
WOOD, 5 TUBES, 2 BANDS
$55

RCA Victor

U104, RADIO-PHONO, c.1938
WOOD, 5 TUBES, 1 BAND
$75

RCA Victor

U111, RADIO-PHONO, c.1938
WOOD, 5 TUBES, 1 BAND
$50

RCA Victor

U112, RADIO-PHONO, c.1938
WOOD, 5 TUBES, 1 BAND
$55

RCA Victor

U115, RADIO-PHONO, c.1938
WOOD, 6 TUBES, 1 BAND
$150

RCA Victor

U119, RADIO-PHONO, c.1938
WOOD, 6 TUBES, 3 BANDS
$100

RCA Victor

V100, RADIO-PHONO, c.1940
WOOD, 5 TUBES, 1 BAND
$60

RCA Victor

V101, RADIO-PHONO, C.1940
WOOD, 6 TUBES, 1 BAND
$50

RCA Victor

V102, RADIO-PHONO, C.1940
WOOD, 7 TUBES, 1 BAND
$75

RCA Victor

VA20, C.1940
JOHN VASSOS DES. (ATTRIB.)
BAKELITE, 2 TUBES, WIRELESS PHONO
$75

RCA Victor

X2-HE, C.1959
PLASTIC, 5 TUBES, 1 BAND
$45

RCA Victor

XF-2, C.1959
PLASTIC, 5 TUBES, 1 BAND
$40

RCA Victor
CANADA

2QX53, C.1955
BAKELITE, 6 TUBES, 3 BANDS
$30

RCA Victor
CANADA

35QU, RADIO-PHONO, C.1954
WOOD, 5 TUBES, 3 BANDS
$45

RCA Victor
CANADA

45E, PHONO, C.1958
BAKELITE, PHONO ONLY
$50

RCA Victor
CANADA

45EY3, PHONO, C.1958
BAKELITE, PHONO ONLY
$60

RCA Victor
CANADA

523, C.1953
BAKELITE, 5 TUBES, 1 BAND
$75

RCA Victor
CANADA

525, C.1955
PLASTIC, 5 TUBES, 1 BAND
$45

RCA Victor
CANADA

533D, C.1958
BAKELITE, 6 TUBES, 3 BANDS
$40

RCA Victor
CANADA
67QB43, c.1959
BAKELITE, 4 TUBES, 3 BANDS
$65

RCA Victor
CANADA
67QR77, c.1959
BAKELITE, 4 TUBES, 6 BANDS
$75

RCA Victor
CANADA
96BT, c.1938
WOOD, 5 TUBES, 1 BAND, DC
$40

RCA Victor
CANADA
'BABY NIPPER', c.1948
NORMAN BEL GEDDES DES.
BAKELITE, 5 TUBES, 1 BAND
$75

RCA Victor
CANADA
BP5D, c.1950
PLASTIC, 4 TUBES, 1 BAND
$40

RCA Victor
CANADA
BT504, c.1958
BAKELITE, 5 TUBES, 1 BAND
$30

RCA Victor
CANADA
C516, c.1954
PLASTIC, 5 TUBES, 1 BAND
$35

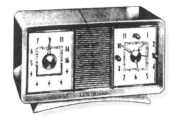

RCA Victor
CANADA
C528, c.1957
PLASTIC, 5 TUBES, 1 BAND
$45

RCA Victor
CANADA
'CHANTICLEER', c.1951
PLASTIC, 5 TUBES, 1 BAND
$45

RCA Victor
CANADA
'NIPPER II', c.1952
PLASTIC, 5 TUBES, 1 BAND
$90

RCA Victor
CANADA
'NIPPER IV, V', c.1956
PLASTIC, 5 TUBES, 1 BAND
$50

RCA Victor
CANADA
'NIPPER VI', c.1957
PLASTIC, 5 TUBES, 1 BAND
$35

RCA Victor
C A N A D A
'NIPPER', C.1951
PLASTIC, 5 TUBES, 1 BAND
$110

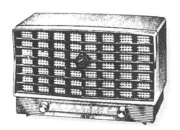

RCA Victor
C A N A D A
X110, C.1957
PLASTIC, 5 TUBES, 1 BAND
$45

RCA Victor
C A N A D A
X310, C.1958
PLASTIC, 5 TUBES, 1 BAND
$125

RCA Victor
C A N A D A
X602, C.1955
PLASTIC, 5 TUBES, 1 BAND
$40

RCI
MIDGET, C.1933
WOOD, 4 TUBES, 1 BAND
$225

Regal
MIDGET, C.1933
WOOD, 4 TUBES, 1 BAND
$325

Regal
1877, C.1954
PLASTIC, 4 TUBES, 1 BAND
$45

Regal
271, C.1954
PLASTIC, 5 TUBES, 1 BAND
$40

Regal
471, C.1954
PLASTIC, 5 TUBES, 1 BAND
$45

Regal
472, C.1954
PLASTIC, 4 TUBES, 1 BAND
$95

Regal
575, C.1954
PLASTIC, 4 TUBES, 1 BAND
$50

Regal
702, C.1954
PLASTIC, 5 TUBES, 1 BAND
$45

Regal

707, c.1954

PLASTIC, 5 TUBES, 1 BAND

$30

Regal

718, c.1954

PLASTIC, 6 TUBES, 2 BANDS

$40

Regal

C473, c.1954

PLASTIC, 4 TUBES, 1 BAND

$110

Regal

C527, c.1954

PLASTIC, 5 TUBES, 1 BAND

$60

REGENT

'4-TUBE', c.1934

WOOD, 4 TUBES, SHORTWAVE

CRYSTAL FINISH $350

REGENT

BSW2W, SW CONV., c.1933

WOOD, 2 TUBES, SHOIRTWAVE

$110

REGENT

PBS4W, c.1934

WOOD, 4 TUBES, SHORTWAVE

CRYSTAL FINISH $350

REGENT

WR4SW, c.1933

4 TUBES, SHORTWAVE, DC

$70

RELIABLE

'MIDGET', c.1933

WOOD, 5 TUBES, 1 BAND

$225

Remington

5T7-055, c.1934

WOOD, 5 TUBES, 1 BAND

$125

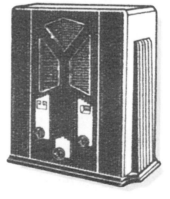

Remington

6T7-055, c.1934

WOOD, 6 TUBES, 2 BANDS

$225

Remington

7T-1322, c.1934

WOOD, 7 TUBES, 3 BANDS

$250

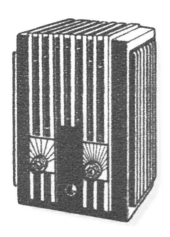

Remington

D7778, C.1934
WOOD, 4 TUBES, 1 BAND
$450

Remington

G5152, G5153, C.1934
WOOD, 5 TUBES, 2 BANDS
$175

Remington

G5163, C.1934
WOOD, 5 TUBES, 2 BANDS
$175

Remington

G5166, G5167, C.1934
WOOD, 6 TUBES, 2 BANDS
$375

Remler

62, 'GRENADIER', C.1936
WOOD, 6 TUBES, 2 BANDS
$250

Remler

14, 'CAMEO', C.1931
WOOD, 1 BAND
$300

Remler

17, C.1931
WOOD, 1 BAND
$350

Remler

60, 'MAYFAIR', C.1938
WOOD, 6 TUBES, 1 BAND
$225

REPUBLIC

111, C.1935
WOOD, 5 TUBES, 1 BAND
$165

REPUBLIC

880, C.1935
WOOD, 4 TUBES, 1 BAND
$190

REPUBLIC

T51, C.1935
WOOD, 5 TUBES, 3 BANDS
$240

REPUBLIC

Y21, C.1935
WOOD, 5 TUBES, 1 BAND
$150

RETS
KIT, C.1946
PLASTIC, 5 TUBES, 1 BAND
$50

Rogers-Majestic
120, 130, 140, C.1949
BAKELITE, 5 TUBES, 1 BAND
$40

Rogers-Majestic
R199, C.1949
WOOD, 13 TUBES, AM/FM/SW
$40

Rogers-Majestic
R202U, C.1959
PLASTIC, 5 TUBES, 1 BAND
WHITE, BROWN, BLACK $75
ORANGE, YELLOW, AQUA $175

Rogers-Majestic
R203U, C.1959
PLASTIC, 5 TUBES, 1 BAND
WHITE, BROWN, BLACK $75
ORANGE, YELLOW, AQUA $175

Rogers-Majestic
R646, C.1932
WOOD, 1 BAND
$160

ROYAL
'DUAL WAVE 5', C.1933
WOOD, 5 TUBES, 2 BANDS
$245

ROYAL
'DUAL WAVE 5', C.1933
WOOD, 5 TUBES, 2 BANDS
$225

Satellite
C.1959
PLASTIC, 5 TUBES, 1 BAND
$45

SCOUT
SW CONVERTER, C.1936
WOOD, 2 TUBES, SHORTWAVE
$90

Sentinel
144XT, C.1939
WOOD, 6 TUBES, 2 BANDS
$65

Sentinel
195U, C.1939
6 TUBES, 2 BANDS
BAKELITE $110
IVORY PLASKON $200
BEETLE $425

Sentinel
298, c.1948

BAKELITE, 5 TUBES, 1 BAND

$65

Sentinel
298T, c.1948

WOOD, 5 TUBES, 1 BAND

$25

Sentinel
302, c.1948

BAKELITE, AM/FM

$90

Sentinel
302T, c.1948

WOOD, AMFM

$75

Sentinel
305, c.1948

BAKELITE, 4 BANDS

$125

Sentinel
309, c.1948

BAKELITE, 5 TUBES, 1 BAND

$50

Sentinel
313, c.1948

BAKELITE, 5 TUBES, 1 BAND

$40

Sentinel
314, c.1948

BAKELITE, 5 TUBES, 1 BAND

$65

Sentinel
315, c.1948

BAKELITE, AM/FM

$40

Sentinel
316P, c.1948

PLASTIC, 4 TUBES, 1 BAND

$50

Sentinel
501, c.1934

WOOD, 5 TUBES, 1 BAND, DC

$100

Sentinel
4230, c.1936

WOOD, 5 TUBES, 1 BAND, DC

$70

Sentinel

4231, c.1936

WOOD, 6 TUBES, 2 BAND, DC

$75

Sentinel

4235, c.1936

WOOD, 5 TUBES, 2 BAND, DC

$75

Sentinel

4236, c.1936

WOOD, 7 TUBES, 3 BAND, DC

$75

Sentinel

SC429, c.1939

WOOD, 5 TUBES, 1 BAND

$125

Setchell-Carlson

411, RADIO/INTERCOM c.1941

DOR-A-FONE, WOOD

$45

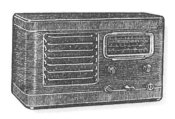

Setchell-Carlson

4140, c.1942

WOOD, 6 TUBES, 4 BANDS

$45

Setchell-Carlson

DUO-PHONE, RADIO/INTERCOM
c.1941

WOOD, 5 TUBES, 1 BAND

$45

SILVER

5E, 'TINY TIM', c.1939

BAKELITE, 5 TUBES, 1 BAND

$100

SILVER

9A4, 'TEENIE WEENIE', c.1939

BAKELITE, 4 TUBES, 1 BAND

$175

SILVER

19W15, c.1939

WOOD, 5 TUBES, 1 BAND

$150

SILVER

100, 'AIRCRUISER', c.1937

WOOD, 5 TUBES, 1 BAND

$750

SILVER

139, c.1937

WOOD, 5 TUBES, 2 BANDS

$350

SILVER

146, c.1937
WOOD, 6 TUBES, 2 BANDS
$300

SILVER

147, c.1937
WOOD, 7 TUBES, 3 BANDS
$100

SILVER

149, c.1937
WOOD, 6 TUBES, 2 BANDS
$375

SILVER

150, c.1937
WOOD, 4 TUBES, 1 BANDS
$115

SILVER

152, c.1937
WOOD, 5 TUBES, 2 BANDS
$125

SILVER

250, c.1937
METAL, 5 TUBES, 1 BAND
$110

SILVER

4801, c.1937
WOOD, 4 TUBES, 2 BANDS, DC
$45

SILVER

501N3, c.1939
WOOD, 5 TUBES, 1 BAND
$60

SILVER

515-5A, c.1939
BAKELITE, 5 TUBES, 1 BAND
$175

SILVER

516-C, c.1939
BAKELITE, 5 TUBES, 1 BAND
$115

SILVER

534, c.1939
WOOD, 8 TUBES, 3 BANDS
$120

SILVER

603, c.1939
WOOD, 8 TUBES, 3 BANDS
$130

SILVER

636V, c.1937

WOOD, 6 TUBES, 3 BANDS, DC

$110

SILVER

6141E, c.1937

WOOD, 7 TUBES, 2 BANDS

$225

SILVER

6322XE, c.1937

WOOD, 8 TUBES, 3 BANDS

$400

SILVER

J6, c.1937

WOOD, 6 TUBES, 3 BANDS

$175

SILVER

M8, c.1937

WOOD, 8 TUBES, 3 BANDS

$325

SILVER

ME17, c.1937

WOOD, 8 TUBES, 3 BANDS

$225

SILVER

P701, c.1939

WOOD, 8 TUBES, 3 BANDS

$225

SILVER-MARSHALL

55A, c.1935

WOOD & CHROME, 5 TUBES, 1 BAND

$200

SILVER-MARSHALL

77T, c.1936

WOOD, 7 TUBES, 3 BANDS

$225

SILVER-MARSHALL

99T, c.1936

WOOD, 9 TUBES, 3 BANDS

$325

SILVER-MARSHALL

714, c.1931

METAL, 6 TUBES, 1 BAND

$75

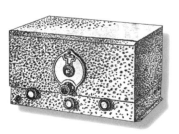

SILVER-MARSHALL

736, SW CONV., c.1931

METAL, 4 TUBES, SHORTWAVE

$75

SILVER-MARSHALL

737, 'BEARCAT', C.1931
METAL, 5 TUBES, SHORTWAVE
$75

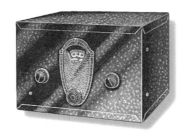

SILVER-MARSHALL

738, SW CONV., C.1931
METAL, 3 TUBES, SHORTWAVE
$75

SILVER-MARSHALL

'ALL WORLD 5', C.1935
WOOD, 5 TUBES, ALLWAVE
$125

SILVER-MARSHALL

T593/693, C.1935
WOOD, 5/6 TUBES, ALLWAVE
$125

Silvertone

4(BROWN), 5(IVORY), C.1960
PLASTIC, 5 TUBES, 1 BAND
$30

Silvertone

7(BROWN), 8(IVORY), 9(BLUE),
10(RED), C.1960
PLASTIC, 5 TUBES, 1 BAND
BROWN, IVORY $30
RED, BLUE $85

Silvertone

11(BROWN), 12(IVORY),
13(BLUE), C.1960
PLASTIC, 6 TUBES, 1 BAND
BROWN, IVORY $45
BLUE $95

Silvertone

15(BROWN), 16(IVORY),
17(GREEN), 18(RED), C.1960
PLASTIC, 6 TRANSISTOR, 1 BAND
BROWN, IVORY $35
RED, GREEN $90

Silvertone

19(BROWN), 20(IVORY),
21(BLUE), 22(RED), C.1960
PLASTIC, 6 TRANSISTOR, 1 BAND
BROWN, IVORY $45
RED, GREEN $95

Silvertone

25(BROWN), 26(IVORY),
C.1960
PLASTIC, 7 TUBES, AM/FM
$50

Silvertone

31, C.1960
PLASTIC, 5 TUBES, 1 BAND
$40

Silvertone

33(BROWN), 34(IVORY),
C.1960
PLASTIC, 5 TUBES, 1 BAND
$40

Silvertone
35(BROWN), 36(IVORY),
37(GREEN), 38(PINK), C.1960
PLASTIC, 5 TUBES, 1 BAND
BROWN, IVORY $50
GREEN, PINK $115

Silvertone
39(BROWN), 40(IVORY),
41(BLUE), C.1960
PLASTIC, 5 TUBES, 1 BAND
BROWN, IVORY $40
BLUE $85

Silvertone
42(BROWN), 43(IVORY),
44(GREEN), 45(PINK) C.1960
PLASTIC, 5 TUBES, 1 BAND
BROWN, IVORY $40
GREEN, PINK $95

Silvertone
215, 216, C.1951
PLASTIC, 5 TUBES, 1 BAND
$70

Silvertone
1571, C.1933
WOOD, 8 TUBES, 2 BANDS, DC
$125

Silvertone
1576, C.1933
WOOD, 8 TUBES, 2 BANDS, DC
$125

Silvertone
1585, C.1933
WOOD, 7 TUBES, 2 BANDS
$300

Silvertone
1589, C.1933
WOOD, 7 TUBES, 2 BANDS
$300

Silvertone
1591, C.1933
WOOD, 6 TUBES, 1 BAND
$225

Silvertone
1601, SW CONV., C.1933
WOOD, 4 TUBES, SHORTWAVE
$110

Silvertone
1621, C.1933
WOOD, 5 TUBES, 1 BAND, DC
$110

Silvertone
1660, C.1933
WOOD, 5 TUBES, 1 BAND
$225

Silvertone
1705, 'WORLD'S FAIR', C.1933
WOOD, 6 TUBES, 1 BAND
$250

Silvertone
1708, 'WORLD'S FAIR', C.1933
WOOD & CHROME
8 TUBES, 2 BANDS
$750

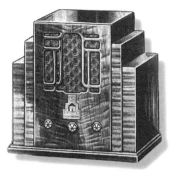

Silvertone
1711, 'WORLD'S FAIR', C.1933
WOOD, 6 TUBES, 1 BAND, DC
$375

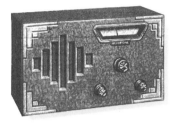

Silvertone
1716, C.1933
5 TUBES, 1 BAND, DC
$75

Silvertone
1724, C.1933
WOOD, 4 TUBES, 1 BAND
$225

Silvertone
1727, C.1933
WOOD, 5 TUBES, 1 BAND
$235

Silvertone
1728, 'LITTLE FELLOW', C.1933
WOOD, 4 TUBES, 1 BAND
$225

Silvertone
1800, C.1935
WOOD, 4 TUBES, 2 BAND
$175

Silvertone
1802, C.1935
WOOD, 5 TUBES, 2 BAND
$135

Silvertone
1804, C.1935
WOOD, 7 TUBES, 2 BAND
$165

Silvertone
1808, 'WORLD'S FAIR', C.1935
WOOD, 6 TUBES, 2 BANDS
$350

Silvertone
1810, C.1935
WOOD, 5 TUBES, 1 BAND
$135

Silvertone

1856, c.1935

WOOD, 8 TUBES, 2 BANDS, DC

$110

Silvertone

1863, c.1935

WOOD, 6 TUBES, 1 BAND, DC

$70

Silvertone

1867, c.1935

WOOD, 8 TUBES, 3 BANDS, DC

$110

Silvertone

1920, c.1936

WOOD, 5 TUBES, 1 BAND, DC

$60

Silvertone

1921, c.1936

WOOD, 6 TUBES, 2 BANDS, DC

$80

Silvertone

1923, c.1936

WOOD, 7 TUBES, 3 BANDS, DC

$125

Silvertone

1953, c.1936

WOOD, 5 TUBES, 2 BANDS

$115

Silvertone

1954, c.1936

WOOD, 6 TUBES, 3 BANDS

$135

Silvertone

1955, c.1936

WOOD, 8 TUBES, 3 BANDS

$225

Silvertone

1980, c.1936

WOOD, 5 TUBES, 1 BAND, DC

$70

Silvertone

1982, c.1936

WOOD, 6 TUBES, 2 BANDS, DC

$125

Silvertone

1983, c.1936

WOOD, 7 TUBES, 3 BANDS, DC

$150

Silvertone
2074, c.1962
PLASTIC, 5 TUBES, 1 BAND
$175

Silvertone
2210, c.1953
PLASTIC, 4 TUBES, 1 BAND
$90

Silvertone
3215(RED), 3217(GREY),
3218(GREEN), c.1953
PLASTIC, 4 TUBES, 1 BAND
GREY $60
RED, GREEN $85

Silvertone
4025(BROWN), 4026(IVORY)
c.1954
PLASTIC, 6 TUBES, 1 BAND
$40

Silvertone
4415(BLACK, 4416(IVORY),
c.1937
BAKELITE, 5 TUBES, 1 BAND
$40

Silvertone
4418, c.1937
WOOD, 4 TUBES, 2 BANDS, DC
$60

Silvertone
4419, c.1937
WOOD, 6 TUBES, 3 BANDS, DC
$65

Silvertone
4420, c.1937
WOOD, 4 TUBES, 1 BANDS, DC
$70

Silvertone
4421, c.1937
WOOD, 4 TUBES, 2 BANDS, DC
$60

Silvertone
4422, c.1937
WOOD, 6 TUBES, 2 BANDS, DC
$100

Silvertone
4424, c.1937
WOOD, 7 TUBES, 3 BANDS, DC
$100

Silvertone
4426, c.1937
WOOD, 9 TUBES, 3 BANDS, DC
$135

Silvertone
4428, c.1937
WOOD, 6 TUBES, 3 BANDS, DC
$65

Silvertone
4429, c.1937
WOOD, 6 TUBES, 2 BANDS, DC
$70

Silvertone
4437, 4438, c.1937
WOOD, 6 TUBES, 2 BANDS
$145

Silvertone
4439, 4440, c.1937
WOOD, 8 TUBES, 3 BANDS, DC
$110

Silvertone
4441, c.1937
WOOD, 8 TUBES, 3 BANDS, DC
$75

Silvertone
4461, c.1937
WOOD, 5 TUBES, 1 BAND
$75

Silvertone
4462, c.1937
WOOD, 5 TUBES, 1 BAND
$110

Silvertone
4463, c.1937
WOOD, 6 TUBES, 3 BANDS
$80

Silvertone
4464, c.1937
WOOD, 5 TUBES, 3 BANDS
$125

Silvertone
4465, c.1937
WOOD, 8 TUBES, 3 BANDS
$250

Silvertone
4469, c.1937
WOOD, 5 TUBES, 2 BANDS
$75

Silvertone
4470, c.1937
WOOD, 6 TUBES, 3 BANDS
$85

Silvertone
4472, 4473, c.1937
WOOD, 4 TUBES, 1 BAND, DC
$110

Silvertone
4498, 4299, c.1937
WOOD, 6 TUBES, 2 BANDS, DC
$110

Silvertone
4500(BLACK), 4505(IVORY), c.1937
BAKELITE/PLASKON, 5 TUBES, 1 BAND
BLACK $125, IVORY $200

Silvertone
4566, c.1937
WOOD, 7 TUBES, 3 BANDS
$115

Silvertone
4569, c.1937
WOOD, 7 TUBES, 3 BANDS
$175

Silvertone
4612, c.1938
WOOD, 6 TUBES, 2 BANDS, DC
$50

Silvertone
4620, c.1938
WOOD, 4 TUBES, 1 BAND, DC
$45

Silvertone
4622, c.1938
WOOD, 4 TUBES, 1 BAND, DC
$55

Silvertone
4623, c.1938
WOOD, 6 TUBES, 2 BANDS, DC
$60

Silvertone
4624, c.1938
WOOD, 6 TUBES, 1 BAND, DC
$75

Silvertone
4626, c.1938
WOOD, 6 TUBES, 2 BANDS, DC
$65

Silvertone
4628, c.1938
WOOD, 8 TUBES, 3 BANDS, DC
$90

Silvertone

4630, c.1938

WOOD, 4 TUBES, 1 BAND, DC

$45

Silvertone

4634, c.1938

WOOD, 6 TUBES, 1 BAND, DC

$45

Silvertone

4636, c.1938

WOOD, 6 TUBES, 2 BANDS, DC

$90

Silvertone

4638, c.1938

WOOD, 8 TUBES, 3 BANDS, DC

$110

Silvertone

4640, c.1938

WOOD, 7 TUBES, 3 BANDS, DC

$80

Silvertone

4660, c.1938

WOOD, 6 TUBES, 2 BANDS

$80

Silvertone

4663, c.1938

WOOD, 7 TUBES, 3 BANDS

$350

Silvertone

4664, c.1938

WOOD, 7 TUBES, 3 BANDS

$165

Silvertone

4665, c.1938

WOOD, 8 TUBES, 3 BANDS

$250

Silvertone

4666, c.1938

WOOD, 9 TUBES, 3 BANDS

$375

Silvertone

4668, RADIO-PHONO, c.1938

WOOD, 6 TUBES, 2 BANDS

$60

Silvertone

4669, c.1938

WOOD, 8 TUBES, 3 BANDS

$325

Silvertone

4728, C.1938
WOOD, 8 TUBES, 3 BANDS, DC
$90

Silvertone

4776, RADIO-PHONO, C.1938
WOOD, 6 TUBES, 2 BANDS, DC
$60

Silvertone

4987, C.1938
WOOD, 8 TUBES, 3 BANDS
$325

Silvertone

5016, 5017, C.1956
WOOD, 5 TUBES, 1 BANDS
$30

Silvertone

5831, PHONO, C.1956
FIBERGLASS, PHONO REMOTE
RED, IVORY $75

Silvertone

6002, C.1946
METAL, 3 TUBES, 1 BAND
$45

Silvertone

6020(BROWN), 6021(IVORY)
C.1956
PLASTIC, 5 TUBES, 1 BAND
$45

Silvertone

6021, C.1939
WOOD, 6 TUBES, 2 BAND
$110

Silvertone

6022, C.1939
WOOD, 7 TUBES, 2 BAND
$140

Silvertone

6023, C.1939
WOOD, 6 TUBES, 2 BANDS
$115

Silvertone

6024, C.1939
WOOD, 8 TUBES, 3 BANDS
$250

Silvertone

6028, RADIO-PHONO, C.1939
WOOD, 6 TUBES, 1 BAND
$50

Silvertone
6040, c.1939
BAKELITE, 4 TUBES, 1 BAND, DC
$60

Silvertone
6042, c.1939
WOOD, 6 TUBES, 1 BAND, DC
$45

Silvertone
6044, c.1939
WOOD, 6 TUBES, 2 BANDS, DC
$60

Silvertone
6046, c.1939
WOOD, 6 TUBES, 2 BANDS, DC
$60

Silvertone
6048, c.1939
WOOD, 7 TUBES, 3 BANDS, DC
$70

Silvertone
6050, RADIO-PHONO, c.1939
WOOD, 4 TUBES, 1 BANDS, DC
$60

Silvertone
6050, c.1946
WOOD & METAL
5 TUBES, 1 BAND
$75

Silvertone
6052, c.1939
WOOD, 6 TUBES, 1 BAND, DC
$100

Silvertone
6070, c.1939
WOOD, 5 TUBES, 1 BAND, DC
$45

Silvertone
6072, c.1939
WOOD, 6 TUBES, 2 BANDS, DC
$80

Silvertone
6073, c.1939
WOOD, 6 TUBES, 2 BANDS
$250

Silvertone
6074, c.1939
WOOD, 7 TUBES, 3 BANDS
$230

Silvertone
6076, c.1939
WOOD, 6 TUBES, 2 BANDS, DC
$65

Silvertone
6102(BLACK), 6103(IVORY)
c.1939
BAKELITE/PLASKON
5 TUBES, 1 BAND
BLACK $125, IVORY $200

Silvertone
6110, 'ROCKET', c.1939
BAKELITE/PLASKON
5 TUBES, 1 BAND
BLACK $1,500
IVORY $2,500

Silvertone
6122, c.1939
WOOD, 7 TUBES, 2 BANDS
$110

Silvertone
6123, c.1939
WOOD, 6 TUBES, 2 BANDS
$85

Silvertone
6124, c.1939
WOOD, 8 TUBES, 3 BANDS
$250

Silvertone
6142, c.1939
WOOD, 6 TUBES, 1 BAND, DC
$40

Silvertone
6146, c.1939
WOOD, 6 TUBES, 2 BANDS, DC
$60

Silvertone
6150, c.1939
WOOD, 4 TUBES, 1 BANDS, DC
$50

Silvertone
6170, c.1939
WOOD, 5 TUBES, 1 BANDS, DC
$40

Silvertone
6172, c.1939
WOOD, 6 TUBES, 2 BANDS, DC
$60

Silvertone
6173, c.1939
WOOD, 6 TUBES, 2 BANDS
$90

Silvertone

6200, c.1939

WOOD, 6 TUBES, 2 BANDS

$90

Silvertone

6201, c.1946

BAKELITE, 4 TUBES, 1 BAND, DC

$20

Silvertone

6202, c.1939

WOOD, 8 TUBES, 2 BANDS

$150

Silvertone

6204, c.1939

WOOD, 6 TUBES, 1 BAND, DC

$45

Silvertone

6206, c.1939

WOOD, 6 TUBES, 2 BANDS, DC

$60

Silvertone

6208, c.1939

WOOD, 7 TUBES, 2 BANDS, DC

$65

Silvertone

6214, c.1939

WOOD, 6 TUBES, 2 BANDS, DC

$100

Silvertone

6221, c.1946

WOOD, 5 TUBES, 1 BANDS, DC

$20

Silvertone

6228, PHONO, c.1940

BAKELITE, PHONO ONLY, DC

$50

Silvertone

6230, c.1939

INGRAHAM, 6 TUBES, 2 BANDS

$125

Silvertone

6231, c.1946

WOOD, 5 TUBES, 4 BANDS, DC

$35

Silvertone

6242, c.1939

WOOD, 5 TUBES, 2 BANDS

$115

Silvertone
6250, c.1939
WOOD, 6 TUBES, 2 BANDS
$75

Silvertone
6251, c.1939
WOOD, 7 TUBES, 2 BANDS
$90

Silvertone
6252, c.1939
WOOD, 8 TUBES, 2 BANDS
$150

Silvertone
6260, RADIO-PHONO, c.1939
WOOD, 4 TUBES, 1 BAND, DC
$50

Silvertone
6261, c.1939
WOOD, 5 TUBES, 1 BAND, DC
$40

Silvertone
6262, c.1939
WOOD, 5 TUBES, 2 BANDS, DC
$60

Silvertone
6263, c.1939
WOOD, 6 TUBES, 2 BANDS, DC
$60

Silvertone
6264, c.1939
WOOD, 6 TUBES, 3 BANDS, DC
$75

Silvertone
6270, c.1939
WOOD, 6 TUBES, 2 BANDS
$250

Silvertone
6271, c.1939
WOOD, 7 TUBES, 2 BANDS
$115

Silvertone
6320, c.1940
WOOD, 6 TUBES, 1 BAND
$225

Silvertone
6321, c.1940
WOOD, 6 TUBES, 2 BAND
$65

Silvertone
6324, c.1940
WOOD, 7 TUBES, 2 BANDS
$75

Silvertone
6325, c.1940
WOOD, 8 TUBES, 3 BANDS
$145

Silvertone
6351, c.1940
WOOD, 4 TUBES, 1 BAND, DC
$35

Silvertone
6354, c.1940
WOOD, 5 TUBES, 1 BAND, DC
$40

Silvertone
6357, c.1940
WOOD, 5 TUBES, 2 BANDS, DC
$40

Silvertone
6360, c.1940
WOOD, 6 TUBES, 4 BANDS, DC
$50

Silvertone
6363, c.1940
WOOD, 6 TUBES, 6 BANDS, DC
$70

Silvertone
6369, c.1940
WOOD, 7 TUBES, 3 BANDS
$100

Silvertone
6372, c.1940
WOOD, 6 TUBES, 1 BAND
$60

Silvertone
6373, c.1940
WOOD, 4 TUBES, 1 BAND, DC
$30

Silvertone
6400(BLACK), 6401(IVORY),
6402(BROWN), c.1940
BAKELITE, 4 TUBES, 1 BAND
IVORY, BROWN $200
BLACK $250
BEETLE (LIMITED PRODUCTION) $750
RED (LIMITED PRODUCTION) $900

Silvertone
6403(BLACK), 6404(IVORY),
6405(BROWN), c.1940
BAKELITE, 5 TUBES, 1 BAND
IVORY, BROWN $125
BLACK $170
BEETLE (LIMITED PRODUCTION) $750

Silvertone

6407(BLACK), 6408(IVORY),
6409(BROWN), C.1940
BAKELITE, 5 TUBES, 1 BAND
IVORY, BROWN $125
BLACK $170
BEETLE (LIMITED PRODUCTION) $750

Silvertone

7000(BROWN), 7002(BEETLE)
C.1942
5 TUBES, 1 BAND
BROWN $125
BEETLE $300

Silvertone

7004(BROWN), 7006(IVORY)
7008(BEETLE), C.1940
5 TUBES, 1 BAND
IVORY, BROWN $75
BEETLE $275

Silvertone

7016, C.1942
BAKELITE, 6 TUBES, 2 BANDS
$40

Silvertone

7025, C.1946
BAKELITE, 5 TUBES, 1 BAND
$75

Silvertone

7036, C.1942
WOOD, 6 TUBES, 2 BANDS
$60

Silvertone

7037, C.1942
WOOD, 7 TUBES, 3 BANDS
$65

Silvertone

7038, C.1942
WOOD, 8 TUBES, 3 BANDS
$90

Silvertone

7039, C.1942
WOOD, 9 TUBES, 4 BANDS
$125

Silvertone

7056, RADIO-PHONO, C.1942
WOOD, 7 TUBES, 1 BAND
$60

Silvertone

7057, RADIO-PHONO, C.1942
WOOD, 7 TUBES, 1 BAND
$60

Silvertone

7058, RADIO-PHONO, C.1942
WOOD, 7 TUBES, 1 BAND
$60

Silvertone

7062, WIRELESS REMOTE, C.1942
BAKELITE, PHONO ONLY
$50

Silvertone

7101, 7102, C.1942
BAKELITE, 4 TUBES, 1 BAND, DC
$20

Silvertone

7105, C.1942
WOOD, 5 TUBES, 1 BAND, DC
$30

Silvertone

7109, C.1942
WOOD, 5 TUBES, 2 BANDS, DC
$40

Silvertone

7113, C.1942
WOOD, 6 TUBES, 4 BANDS, DC
$40

Silvertone

7120, 7123, C.1942
WOOD, 6 TUBES, 1 BAND, DC
$45

Silvertone

7127, C.1942
WOOD, 7 TUBES, 3 BANDS
$75

Silvertone

8000, C.1949
BAKELITE, 4 TUBES, 1 BAND
$85

Silvertone

8003(GREY), 8004(GREEN)
C.1949
METAL, 4 TUBES, 1 BAND
$50

Silvertone

8005, C.1949
BAKELITE, 5 TUBES, 1 BAND
$60

Silvertone

8020, C.1949
BAKELITE, 8 TUBES, AM/FM
$60

Silvertone

8022, C.1949
BAKELITE, 7 TUBES, AM/FM
$50

Silvertone

8051, C.1949

BAKELITE, 5 TUBES, 1 BAND

$30

Silvertone

8052, C.1949

BAKELITE, 5 TUBES, 1 BAND

$30

Silvertone

8070, RADIO-PHONO, C.1949

BAKELITE, 4 TUBES, 1 BAND

$60

Silvertone

8080, RADIO-PHONO, C.1949

BAKELITE, 4 TUBES, 1 BAND

$50

Silvertone

8155, RADIO REMOTE, C.1949

BAKELITE, 1 TUBES, PHONO ONLY

$60

Silvertone

8203, C.1949

BAKELITE, 4 TUBES, 1 BAND, DC

$20

Silvertone

8213, RADIO-PHONO, C.1949

WOOD, 4 TUBES, 1 BAND, DC

$25

Silvertone

8222, C.1949

WOOD, 5 TUBES, 1 BAND, DC

$25

Silvertone

8231, C.1949

WOOD, 6 TUBES, 4 BANDS, DC

$25

Silvertone

8270, C.1949

BAKELITE, 5 TUBES, 1 BAND

$50

Silvertone

9018, 'METEOR, C.1960

PLASTIC, 4 TUBES, 1 BAND

$45

Silvertone

9260, C.1949

PLASTIC, 4 TUBES, 1 BAND

$75

Silvertone

'MISSION BELL', C.1937
WOOD, 5 TUBES, 1 BAND
$650

Silvertone
CANADA

8027, C.1958
PLASTIC, 5 TUBES, 1 BAND
PINK $75

Simplex

C.1936
WOOD, 5 TUBES, 2 BANDS, DC
$90

Simplex

7D, C.1939
WOOD, 7 TUBES, 2 BANDS
$110

Simplex

D, 'DELUXE', C.1936
WOOD, 7 TUBES, 2 BANDS
$350

Simplex

NT, C.1939
WOOD, 11 TUBES, 3 BANDS
$325

Simplex

P, C.1934
WOOD, 5 TUBES, 2 BANDS
$225

Simplex

R, 'JUNIOR', C.1934
WOOD, 4 TUBES, 2 BANDS
$250

Simplex

'RADIOPHONE', C.1937
WOOD, INTERCOM SET
$75

Simplex

RJH, C.1939
WOOD, 6 TUBES, 2 BANDS
$75

Simplex

RK, C.1936
WOOD, 4 TUBES, 1 BAND
$150

Simplex

'SPORTSMAN', C.1936
CANVAS, 4 TUBES, 1 BAND
$50

Sonic

'Pee Wee', c.1956
PLASTIC, 4 TUBES, 1 BAND
$115

SONOLA

'Midget', c.1931
WOOD, 5 TUBES, 1 BAND
$225

Sonora

c.1942
WOOD, 4 TUBES, 1 BAND, DC
$30

Sonora

26, c.1940
WOOD, 5 TUBES, 1 BAND
$55

Sonora

38, 'Duet', c.1940
WOOD, PHONO ONLY
$90

Sonora

40, 'Trio', c.1940
WOOD, PHONO ONLY
$150

Sonora

41, 'Sonophonic', c.1940
WOOD, PHONO ONLY
$90

Sonora

46, c.1940
WOOD, 4 TUBES, 1 BAND, DC
$30

Sonora

165-6W, c.1938
WOOD, 6 TUBES, 2 BANDS
$70

Sonora

170-6B, c.1938
WOOD, 6 TUBES, 3 BANDS, DC
$40

Sonora

175, c.1938
WOOD, 8 TUBES, 3 BANDS
$165

Sonora

205-6Q, c.1938
WOOD, 6 TUBES, 2 BANDS, DC
$60

Sonora

209, c.1945
INGRAHAM, 5 TUBES, 1 BAND
$125

Sonora

216-6B, c.1945
WOOD, 6 TUBES, 3 BANDS, DC
$50

Sonora

KE-78, RADIO-PHONO, c.1940
BAKELITE, 5 TUBES, 1 BAND
$70

Sonora

KM-138, c.1941
CATALIN, 5 TUBES, 1 BAND
$700+

Sonora

TE-38, c.1940
WOOD, PHONO ONLY
$90

Sonora

TE-79, c.1940
WOOD, PHONO ONLY
$165

Sonora

TNF-60, RADIO-PHONO, c.1940
WOOD, 5 TUBES, 1 BAND
$70

Sparton

5G1-K, c.1958
PLASTIC, 5 TUBES, 1 BAND
$60

Sparton

5HK-1, c.1958
PLASTIC, 5 TUBES, 1 BAND
$60

Sparton

315C, 'FOOTBALL', c.1953
PLASTIC, 5 TUBES, 1 BAND
$125

Sparton

325T, c.1955
PLASTIC, 5 TUBES, 1 BAND
$90

Sparton

417X, c.1937
WOOD, 4 TUBES, 1 BAND
$350

Sparton

510DG (LTD. ED.), C.1939
INGRAHAM, 5 TUBES, 1 BAND
$250

Sparton

510DR (LTD. ED.), C.1939
INGRAHAM, 5 TUBES, 1 BAND
$250

Sparton

510W (LTD. ED.), C.1939
INGRAHAM, 5 TUBES, 1 BAND
$250

Sparton

527-2, C.1937
WOOD, 5 TUBES, 2 BANDS, DC
$100

Sparton

617, C.1937
WOOD, 6 TUBES, 2 BANDS
$175

Sparton

727X, 827X, C.1937
WOOD, 7,8 TUBES, 2 BANDS
7 TUBES $225, 8 TUBES $275

Sparton

'MORNING STAR', C.1954
PLASTIC, 5 TUBES, 1 BAND
$90

Sparton

'TABLE TOPPER', C.1954
PLASTIC, 5 TUBES, 1 BAND
$90

Sparton
CANADA

253, C.1935
WOOD, 5 TUBES, 1 BAND
$250

Sparton
CANADA

4099, 'FOOTBALL', C.1950
PLASTIC, 5 TUBES, 1 BAND
$125

Sparton
CANADA

4641K, C.1940
WOOD, 5 TUBES, 1 BAND, DC
$60

Sparton
CANADA

5048L, C.1948
5 TUBES, 1 BAND
$35

Sparton
CANADA
5442K, 'PARK AVE.', C.1940
7 TUBES, 2 BANDS
$150

Sparton
CANADA
6049K, C.1948
6 TUBES, 1 BAND
$35

Sparton
CANADA
6049KC, RADIO-PHONO, C.1948
6 TUBES, 1 BAND
$50

Sparton
CANADA
6149K, C.1948
6 TUBES, 1 BAND
$40

Sparton
CANADA
7241K, C.1940
7 TUBES, 2 BANDS
$275

Sparton
CANADA
8549XK, C.1948
8 TUBES, 5 BANDS
$110

STAR–LITE
AF3, C.1960
PLASTIC, 7 TUBES, AMFM
$75

STEINITE
600, C.1931
WOOD, 6 TUBES, 1 BAND
$275

STEINITE
605, RADIO-PHONO, C.1931
WOOD, 6 TUBES, 1 BAND
$350

STEINITE
'JUNIOR', C.1931
WOOD, 5 TUBES, 1 BAND
$250

STEWART-WARNER
2-421, C.1940
WOOD, 4 TUBES, 1 BAND, DC
$30

STEWART-WARNER
2A1, WIRELESS PHONO, C.1941
WOOD, PHONO ONLY
$45

STEWART-WARNER

3-5K3, c.1940

BAKELITE/PLASKON
5 TUBES, 1 BAND
BROWN $150, IVORY $250

STEWART-WARNER

4B1, c.1941

BAKELITE, 4 TUBES, 1 BAND, DC
$70

STEWART-WARNER

4C1, c.1941

WOOD, 4 TUBES, 1 BAND, DC
$40

STEWART-WARNER

4C2, c.1942

WOOD, 4 TUBES, 1 BAND, DC
$35

STEWART-WARNER

5T1, c.1941

WOOD, 5 TUBES, 2 BANDS, DC
$45

STEWART-WARNER

5V9, RADIO-PHONO, c.1941

WOOD, 5 TUBES, 1 BAND
$50

STEWART-WARNER

205AA9(BROWN),
205AB(IVORY), c.1942

BAKELITE, 5 TUBES, 1 BAND
$40

STEWART-WARNER

205AC, c.1942

WOOD, 5 TUBES, 1 BAND
$40

STEWART-WARNER

205CA, c.1942

WOOD, 5 TUBES, 2 BANDS, DC
$30

STEWART-WARNER

205FA, RADIO-PHONO, c.1942

WOOD, 5 TUBES, 1 BANDS
$45

STEWART-WARNER

206BA9(BROWN),
206BB(IVORY), c.1942

BAKELITE, 5 TUBES, 1 BAND
$45

STEWART-WARNER

205BC, c.1942

WOOD, 6 TUBES, 1 BAND
$40

STEWART-WARNER

206DA9(BROWN),
206DB(IVORY), c.1942
BAKELITE, 6 TUBES, 2 BANDS
$45

STEWART-WARNER

206DC, c.1942
WOOD, 6 TUBES, 2 BANDS
$45

STEWART-WARNER

207BA, c.1942
WOOD, 7 TUBES, 3 BANDS
$50

STEWART-WARNER

511(BROWN), 513(IVORY), 'CAMPUS',
c.1940
BAKELITE/PLASKON
5 TUBES, 1 BAND
BROWN $150, IVORY $250

STEWART-WARNER

1111, 'WORLD'S FAIR', c.1933
WOOD & METAL, 4 TUBES, 1 BAND
GREEN, BURLE $1,500

STEWART-WARNER

1163, c.1934
WOOD, 4 TUBES, 1 BAND
$325

STEWART-WARNER

1302, c.1936
WOOD, 5 TUBES, 2 BANDS
$225

STEWART-WARNER

1361, c.1936
WOOD, 7 TUBES, 3 BANDS
$275

STEWART-WARNER

1362, c.1936
WOOD, 7 TUBES, 3 BANDS
$325

STEWART-WARNER

1401, c.1936
WOOD, 5 TUBES, 2 BANDS
$150

STEWART-WARNER

1801, c.1938
WOOD, 5 TUBES, 2 BANDS
$100

STEWART-WARNER

1802, c.1938
WOOD, 5 TUBES, 2 BANDS
$120

STEWART-WARNER

1812, C.1938
WOOD, 6 TUBES, 3 BANDS
$135

STEWART-WARNER

1881, C.1938
WOOD, 6 TUBES, 2 BANDS
$80

STEWART-WARNER

3041, C.1938
WOOD, 5 TUBES, 1 BAND
$750

STEWART-WARNER

3042, C.1938
WOOD, 5 TUBES, 1 BAND
$300

STEWART-WARNER

A6, 'AIR PAL', C.1941
BAKELITE, 6 TUBES, 1 BAND
$175

STEWART-WARNER

B61T1(IVORY), B61T2(BROWN),
C.1950
BAKELITE, 6 TUBES, 1 BAND
$85

STEWART-WARNER

C51T1, C.1950
BAKELITE, 5 TUBES, 1 BAND
$60

STEWART-WARNER

'PORTO-BARADIO', C.1942
BAKELITE, 5 TUBES, 1 BAND
$175 (COMPLETE)

STEWART-WARNER
C A N A D A
4591, REMOTE TUNER, C.1940
WOOD, 3 TUBES, REMOTE
$250

STEWART-WARNER
C A N A D A
4721, C.1941
WOOD, 4 TUBES, 1 BAND, DC
$30

STEWART-WARNER
C A N A D A
4741, C.1941
WOOD, 6 TUBES, 2 BANDS
$50

STEWART-WARNER
C A N A D A
4791, C.1941
WOOD, 5 TUBES, 2 BANDS
$40

STEWART-WARNER
C A N A D A
4831, c.1941
WOOD, 5 TUBES, 2 BANDS, DC
$35

STEWART-WARNER
C A N A D A
4862, c.1941
BAKELITE/PLASKON
5 TUBES, 1 BAND
BROWN $150, IVORY $250

STEWART-WARNER
C A N A D A
9181, c.1956
PLASTIC, 5 TUBES, 1 BAND
$45

STEWART-WARNER
C A N A D A
9187, c.1956
PLASTIC, 5 TUBES, 1 BAND
$65

STEWART-WARNER
C A N A D A
'COMET', c.1938
WOOD, 5 TUBES, 1 BAND
$45

STEWART-WARNER
C A N A D A
M50, c.1949
WOOD, 7 TUBES, 2 BANDS
$55

STEWART-WARNER
C A N A D A
R603-9, c.1951
WOOD, 4 TUBES, 1 BAND
$30

stratford
'DELUXE', c.1934
WOOD, 4 TUBES, 1 BAND
$225

stratford
'TRF', c.1934
WOOD, 4 TUBES, 1 BAND
$225

stratford
'4-TUBE', c.1934
WOOD, 4 TUBES, 1 BAND
$150

stratford
'SUPERHET', c.1932
WOOD, 5 TUBES, 1 BAND
$110

stratford
'SUPERHET', c.1934
WOOD, 7 TUBES, 1 BAND
$200

stratford

'8-TUBE', C.1934
WOOD, 8 TUBES, 1 BAND
$250

stratford

'ALL-WAVE 7', C.1933
WOOD, 7 TUBES, 1 BAND
$225

stratford

'DELUXE 4', C.1932
WOOD, 4 TUBES, 1 BAND
$275

stratford

'DELUXE 7', C.1932
WOOD, 7 TUBES, 1 BAND
$235

stratford

'DELUXE COMPACT', C.1933
WOOD, 5 TUBES, 1 BAND
$125

stratford

'DELUXE MODERNE', C.1934
WOOD, 8 TUBES, 3 BANDS
$350

stratford

'DELUXE', C.1934
REPWOOD, 5 TUBES, 1 BAND
$375

stratford

'DELUXE', C.1934
METAL, 5 TUBES, SHORTWAVE
$75

stratford

'DUAL WAVE', C.1934
WOOD, 5 TUBES, 2 BANDS
$225

stratford

DW4, C.1934
WOOD, 4 TUBES, 2 BANDS
$225

stratford

DW5, 'MODERNE', C.1934
WOOD, 5 TUBES, 2 BANDS
$325

stratford

'INGENUE', C.1933
WOOD, 5 TUBES, 1 BAND
$225

Stratford
'MIGHTY MIDGET', C.1932
WOOD, 5 TUBES, 1 BAND
$125

Stratford
'MODERNE', C.1934
WOOD, 5 TUBES, 1 BAND
$235

Stratford
'ROUND THE WORLD', C.1934
WOOD, 5 TUBES, 2 BANDS
$150

Stratford
'SCREEN GRID', C.1931
WOOD, 6 TUBES, 1 BAND
$145

Stratford
'SENIOR 7', C.1932
WOOD, 6 TUBES, 1 BAND
$245

Stratford
'UNIVERSAL', C.1934
WOOD, 5 TUBES, 1 BAND
$225

STROMBERG
550AW, C.1932
WOOD, 5 TUBES, 2 BANDS
$225

Stromberg-Carlson
130J, 'TREASURE CHEST', C.1937
WOOD, 7 TUBES, 2 BANDS
$350

Stromberg-Carlson
235H, C.1938
WOOD, 8 TUBES, 3 BANDS
$300

Stromberg-Carlson
425H, C.1940
WOOD, 8 TUBES, AMFM
$225

Stromberg-Carlson
430H, C.1939
WOOD, 9 TUBES, 2 BANDS
$275

Stromberg-Carlson
1202, 'MODERNAIRE'
RADIO-PHONO, C.1949
WOOD, 5 TUBES, 5 BAND
$125

Stromberg-Carlson

2152H, c.1948
WOOD, 7 TUBES, 2 BANDS
$75

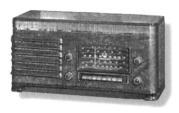

Stromberg-Carlson
C A N A D A

75H, c.1941
WOOD, 9 TUBES, 5 BANDS
$175

Stromberg-Carlson
C A N A D A

874H, c.1948
WOOD, 7 TUBES, 4 BANDS
$125

Stromberg-Carlson
C A N A D A

961HW, c.1949
BAKELITE, 6 TUBES, 1 BAND
$75

Stromberg-Carlson
C A N A D A

1141, c.1948
BAKELITE, 4 TUBES, 1 BAND, DC
$45

Stromberg-Carlson
C A N A D A

1151, c.1948
BAKELITE, 5 TUBES, 1 BAND
$150

Stromberg-Carlson
C A N A D A

1151A, c.1951
BAKELITE, 5 TUBES, 1 BAND
$150

Stromberg-Carlson
C A N A D A

4151, c.1948
BAKELITE, 5 TUBES, 1 BAND
$75

SUBMARINER

SW ADAPTER, c.1931
WOOD, SHORTWAVE
$75

SYLVANIA

518, c.1955
PLASTIC, 5 TUBES, 1 BAND
RED, GREEN $90
BLACK, TAN $40

SYLVANIA

543, c.1955
PLASTIC, 5 TUBES, 1 BAND
$45

SYLVANIA

548, c.1955
PLASTIC, 5 TUBES, 1 BAND
RED, GREEN $70
BLACK, TAN $35

SYLVANIA

598, c.1955
PLASTIC, 5 TUBES, 1 BAND
RED, GREEN $90
BLACK, TAN $40

SYLVANIA

614, c.1955
PLASTIC, 5 TUBES, 1 BAND
RED, GREEN $95
BLACK, TAN $40

SYLVANIA

619, c.1955
PLASTIC, 5 TUBES, 1 BAND
RED, GREEN $90
BLACK, IVORY $40

SYLVANIA

5151, c.1955
PLASTIC, 5 TUBES, 1 BAND
RED, GREEN $80
BLACK, IVORY $40

SYLVANIA

5184, c.1955
PLASTIC, 5 TUBES, 1 BAND
RED, GREEN $70
BLACK, IVORY $35

SYLVANIA

'CATALINA', c.1953
PLASTIC
$65

SYLVANIA

Z5T17', c.1958
PLASTIC
$70

TEE-NEE

'ONE', c.1941
WOOD, CRYSTAL
$175

TEE-NEE

c.1941
PLASTIC, CRYSTAL
$225

Tele-tone

'MIDGET', c.1932
WOOD, 7 TUBES, 1 BAND
$245

Tele-tone

TR155, c.1958
PLASTIC, 5 TUBES, 1 BAND
$90

Temple RADIO

G4, c.1948
METAL, 4 TUBES, 1 BAND
$75

Temple
G522, C.1949
WOOD, 5 TUBES, 1 BAND
$35

TIFFANY-TONE
5A, C.1936
WOOD, 5 TUBES, 1 BAND
$150

TIFFANY-TONE
5MT, C.1936
WOOD, 5 TUBES, 1 BAND
$250

TIFFANY-TONE
5MW, C.1936
WOOD & MIRROR
5 TUBES, 1 BAND
$1,200

TIFFANY-TONE
7MT, C.1936
WOOD, 7 TUBES, 2 BANDS
$275

TIFFANY-TONE
9MT, C.1936
WOOD, 9 TUBES, 3 BANDS
$950

TIFFANY-TONE
62, C.1931
WOOD, 6 TUBES, 1 BAND
$350

TIFFANY-TONE
66MT, 66GT, C.1936
WOOD, 6 TUBES, 2 BANDS
$300

TIFFANY-TONE
156A, C.1935
WOOD, 5 TUBES, 1 BAND
$235

TIFFANY-TONE
518PR, RADIO-PHONO, C.1938
WOOD, 5 TUBES, 1 BAND
$225

Tiny Tim
526-4H, C.1940
BAKELITE, 4 TUBES, 1 BAND
$65

TRAV-LER
C.1938
BAKELITE, 5 TUBES, 1 BAND
$70

TRAV-LER

30, C.1951

PLASTIC, 6 TUBES, 1 BAND

$45

TRAV-LER

36, C.1951

BAKELITE, 6 TUBES, 1 BAND

$45

TRAV-LER

55C42, C.1955

PLASTIC, 5 TUBES, 1 BAND

$40

TRAV-LER

66, C.1951

BAKELITE, 7 TUBES, 1 BAND

$35

TRAV-LER

336, C.1940

WOOD, 6 TUBES, 3 BANDS

$80

TRAV-LER

400, C.1940

WOOD, 4 TUBES, 1 BAND

$165

TRAV-LER

400W, C.1940

WOOD, 4 TUBES, 1 BAND

$165

TRAV-LER

415, C.1940

WOOD, 5 TUBES, 1 BAND

$90

TRAV-LER

415W, C.1940

WOOD, 5 TUBES, 1 BAND

$90

TRAV-LER

425M, C.1940

WOOD, 5 TUBES, 1 BAND

$110

TRAV-LER

436M, C.1940

WOOD, 6 TUBES, 3 BANDS

$125

TRAV-LER

455L, RADIO-PHONO, C.1939

WOOD, 5 TUBES, 2 BANDS

$55

TRAV-LER
465M, c.1939
WOOD, 5 TUBES, 2 BANDS
$95

TRAV-LER
536M, c.1939
WOOD, 7 TUBES, 3 BANDS
$125

TRAV-LER
539M, c.1939
WOOD, 9 TUBES, 4 BANDS
$165

TRAV-LER
552B, c.1939
WOOD, 4 TUBES, 1 BAND, DC
$45

TRAV-LER
560B, c.1939
WOOD, 5 TUBES, 1 BAND, DC
$50

TRAV-LER
575B, c.1939
WOOD, 6 TUBES, 3 BANDS, DC
$55

TRAV-LER
630, c.1938
WOOD, 6 TUBES, 2 BANDS
$90

TRAV-LER
720, c.1940
WOOD, 7 TUBES, 1 BAND
$85

TRAV-LER
730, c.1938
WOOD, 7 TUBES, 3 BANDS
$125

TRAV-LER
802, c.1940
WOOD, 7 TUBES, 1 BAND
$75

TRAV-LER
830, c.1938
WOOD, 8 TUBES, 3 BANDS
$150

TRAV-LER
3501, RADIO-PHONO, c.1940
WOOD, 5 TUBES, 1 BAND
$50

TRAV-LER

3503, c.1940

WOOD, 7 TUBES, 4 BANDS

$130

TRAV-LER

3522, c.1940

WOOD, 7 TUBES, 2 BANDS

$100

TRAV-LER

3524, c.1940

WOOD, 7 TUBES, 3 BANDS

$115

TRAV-LER

5061, c.1950

BAKELITE, 5 TUBES, 1 BAND

$40

TRAV-LER

5170, c.1950

WOOD, 5 TUBES, 1 BAND

$30

Troubador

2561, c.1947

BAKELITE, 5 TUBES, 1 BAND

$30

TROY

4, c.1937

WOOD, 4 TUBES, 1 BAND

$85

TROY

53, c.1937

WOOD, 5 TUBES, 3 BANDS

$225

TROY

75, c.1937

WOOD, 5 TUBES, 1 BAND

$275

TROY

75PC, RADIO-PHONO, c.1937

WOOD, 5 TUBES, 1 BAND

$225

TROY

77C, c.1937

WOOD, 7 TUBES, 3 BANDS

$275

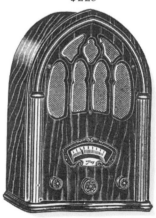

TROY

'MIGHTY MIDGET', c.1933

WOOD, 5 TUBES, 1 BAND

$225

TRUETONE

'COMBINATION', RADIO-PHONO,
C.1939
WOOD, 7 TUBES, 2 BANDS
$55

TRUETONE

D709, C.1939
WOOD, 4 TUBES, 1 BAND, DC
$30

TRUETONE

D711, C.1938
WOOD, 8 TUBES, 3 BANDS
$210

TRUETONE

D713, C.1938
WOOD, 6 TUBES, 2 BANDS
$70

TRUETONE

D715, C.1938
WOOD, 5 TUBES, 1 BAND
$45

TRUETONE

D716, C.1938
WOOD, 7 TUBES, 2 BANDS
$90

TRUETONE

D719, C.1938
WOOD, 4 TUBES, 1 BAND, DC
$30

TRUETONE

D719A, C.1938
WOOD, 6 TUBES, 2 BAND
$50

TRUETONE

D730, C.1938
BAKELITE, 5 TUBES, 1 BAND
$95

TRUETONE

D731, C.1938
BAKELITE, 6 TUBES, 1 BAND
$125

TRUETONE

D903, PHONO REMOTE,
C.1940
WOOD, PHONO ONLY
$65

TRUETONE

D910, C.1940
WOOD, 6 TUBES, 2 BANDS
$100

TRUETONE

D911, c.1940

WOOD, 8 TUBES, 3 BANDS

$115

TRUETONE

D914, c.1940

WOOD, 7 TUBES, 3 BANDS

$120

TRUETONE

D935, c.1940

WOOD, 4 TUBES, 1 BAND, DC

$25

TRUETONE

D1001, c.1941

INGRAHAM, 5 TUBES, 1 BAND

$150

TRUETONE

D1019, c.1942

BAKELITE, 5 TUBES, 1 BAND

$30

TRUETONE

D1104, c.1942

WOOD, 7 TUBES, 2 BANDS

$40

TRUETONE

D1118, c.1942

BAKELITE, 7 TUBES, 2 BANDS

$50

TRUETONE

D1120, c.1942

WOOD, 4 TUBES, 1 BAND, DC

$25

TRUETONE

D1123, c.1942

WOOD, 6 TUBES, 3 BANDS, DC

$35

TRUETONE

D1124, c.1942

BAKELITE, 4 TUBES, 1 BAND, DC

$75

TRUETONE

D1135, c.1942

WOOD, 4 TUBES, 1 BAND, DC

$25

TRUETONE

D1170, RADIO-PHONO, c.1942

WOOD, 5 TUBES, 1 BAND

$45

TRUETONE

D1171, RADIO-PHONO, C.1942
WOOD, 5 TUBES, 1 BANDS
$45

TRUETONE

D1172, RADIO-PHONO, C.1942
WOOD, 6 TUBES, 2 BANDS
$75

TRUETONE

D2389, C.1953
BAKELITE, 5 TUBES, 1 BAND
$45

TRUETONE

D2603, PHONO, C.1947
WOOD, 3 TUBES, PHONO ONLY
$45

TRUETONE

D2610, C.1947
BAKELITE, 5 TUBES, 1 BAND
$190

TRUETONE

D2615, C.1947
BAKELITE, 6 TUBES, 1 BAND
$100

TRUETONE

D2616A, C.1947
BAKELITE, 6 TUBES, 1 BAND
$90

TRUETONE

D2619, C.1947
WOOD, 6 TUBES, 1 BAND
$30

TRUETONE

D2640, RADIO-PHONO, C.1947
WOOD, 5 TUBES, 1 BAND
$45

TRUETONE

D2743, RADIO-PHONO, C.1947
WOOD, 4 TUBES, 1 BAND
$60

TRUETONE

D3809, C.1946
PLASTIC, 4 TUBES, 1 BAND, DC
$45

TRUETONE

D3811, C.1948
PLASTIC & METAL
4 TUBES, 1 BAND, DC
$60

TRUETONE

DC2036, c.1960

PLASTIC, 5 TUBES, 1 BAND

$60

TRUETONE

DC2170, c.1960

PLASTIC, 4 TUBES, 1 BAND

$25

TRUETONE

DC2261, c.1960

PLASTIC, 4 TUBES, 1 BAND

$25

TRUETONE

DC2284, c.1961

PLASTIC, 6 TUBES, 1 BAND

$30

TRUETONE

DC2287, c.1961

PLASTIC, 6 TUBES, 1 BAND

$40

TRUETONE

DC2362, c.1961

PLASTIC, 5 TUBES, 1 BAND

$30

TRUETONE

DC2371, c.1961

PLASTIC, 5 TUBES, 1 BAND

$35

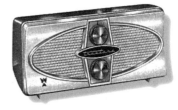

TRUETONE

DC2380, c.1961

PLASTIC, 5 TUBES, 1 BAND

$75

TRUETONE

DC2989, c.1961

PLASTIC, 5 TUBES, 1 BAND

$25

TRUETONE

DD709, c.1938

WOOD, 6 TUBES, 2 BANDS, DC

$25

TRUETONE

'ENVOY', c.1939

WOOD, 6 TUBES, 2 BANDS

$75

Universal

50A6, c.1937

WOOD, 5 TUBES, 1 BANDS

$135

Universal

5010, C.1936
WOOD, 5 TUBES, 1 BAND
$65

Universal

6110, C.1936
WOOD, 6 TUBES, 2 BANDS
$250

Universal

8210, C.1936
WOOD, 8 TUBES, 3 BANDS
$300

UNKNOWN

C.1934
WOOD & METAL
$350

U S

3092, C.1933
WOOD, 4 TUBES, 1 BAND
$110

VIKING

4, C.1933
WOOD, 4 TUBES, 1 BAND
$175

VIKING

5, C.1933
WOOD, 5 TUBES, 1 BAND
$210

Walker

4X, SW CONV., C.1932
METAL, 4 TUBES, SHORTWAVE
$110

Walker

MULTI-UNIT, C.1931
1 TUBE, BC/SW
$150

Walker

'SPECIAL', SW ADAPTER, C.1931
METAL, 1 TUBE, SHORTWAVE
$75

WALTON

C.1933
WOOD, 4 TUBES, 1 BAND
$145

WARWICK

C.1934
WOOD & ALUMINUM
5 TUBES, 1 BAND
$275

247

Wells-Gardner

108A1-74, C.1938
WOOD, AW
$145

Wells-Gardner

7D14, C.1947
BAKELITE, 7 TUBES, 2 BANDS
$65

Wells-Gardner

7D14W, C.1947
WOOD, 7 TUBES, 3 BANDS
$60

WESTERN

66-14913, C.1935
WOOD, 7 TUBES, 2 BANDS
$375

Westinghouse

434T5(BLACK), 435T5(IVORY),
436T5(RED), 437T5(TAN),
438T5(GREEN), 440T5(GREY), C.1957
PLASTIC, 5 TUBES, 1 BANDS
BLACK, IVORY, TAN, GREY $60
RED, GREEN $90

Westinghouse

435T5A(IVORY), 437T5A(TAN),
438T5A(GREEN), C.1957
PLASTIC, 5 TUBES, 1 BANDS
IVORY, TAN $60
GREEN $90

Westinghouse

486T5(IVORY), 487T5(RED),
488T5(BLACK), 489T5(GREY), C.1957
PLASTIC, 5 TUBES, 1 BANDS
IVORY, BLACK, GREY $50
RED $75

Westinghouse

499T5(BLACK), 500T5(RED),
501T5(TAN), 551T5(GREEN)IVORY,
BLACK, TAN $45
RED, GREEN $75

Westinghouse

503T5, C.1957
PLASTIC, 5 TUBES, 1 BAND
GREY & PLAID $65

Westinghouse

519P4(MAROON), 520P4(RED),
521P4(TAN), C.1955
PLASTIC, 4 TUBES, 1 BAND
MAROON, TAN $60
RED $90

Westinghouse

536T6, C.1957
PLASTIC, 6 TUBES, 1 BAND
$45

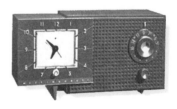

Westinghouse

538T4BLACK), 539T4(IVORY),
540T4(RED), C.1955
PLASTIC, 4 TUBES, 1 BAND
BLACK, IVORY $75
RED $145

Westinghouse

541T5(GREY), 542T5(BROWN),
543T5(IVORY), C.1957
PLASTIC, 5 TUBES, 1 BAND
$50

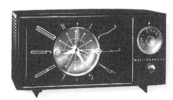

Westinghouse

544T5(IVORY), 545T5(PINK),
546T5(BLACK), C.1957
PLASTIC, 5 TUBES, 1 BAND
IVORY, BLACK $50
PINK $75

Westinghouse

547T5(RED), 548T5(GREEN),
548T5(BLACK), 550T5(PINK), C.1957
PLASTIC, 5 TUBES, 1 BAND
BLACK $70
RED, GREEN, PINK $100

Westinghouse

557P4(GREEN), 558P4(WHITE/TAN),
559P4(GREY/BLACK), C.1957
PLASTIC, 4 TUBES, 1 BAND
$80

Westinghouse

562P4(TAN/BROWN),
563P4(GREY/BLACK),
564P4(WHITE/TAN), C.1957
PLASTIC, 4 TUBES, 1 BAND
$75

Westinghouse

570T4(BROWN), 571T4(IVORY),
572T4(PINK), C.1957
PLASTIC, 4 TUBES, 1 BAND
BROWN, IVORY $50
PINK $75

Westinghouse

574T4(BLACK), 575T4(IVORY),
576T4(PINK), 577T4(RED), C.1957
PLASTIC, 4 TUBES, 1 BAND
BLACK, IVORY $45
PINK $60
RED $90

Westinghouse

580T5(BROWN), 581T5(IVORY),
583T5(RED), C.1957
PLASTIC, 4 TUBES, 1 BAND
BROWN, IVORY $85
RED $160

Westinghouse

621PX(GREY), 622PX(YELLOW),
C.1957
PLASTIC, 6 TUBES, 1 BAND
GREY $85
YELLOW $160

Westinghouse

626T5(IVORY), 627T5(YELLOW),
628T5(AQUA), C.1957
PLASTIC, 5 TUBES, 1 BAND
IVORY $45
YELLOW, AQUA $80

Westinghouse

629T4(IVORY), 630T4(PINK),
631T4(GREEN), C.1957
PLASTIC, 4 TUBES, 1 BAND
IVORY $45
PINK, GREEN $60

Westinghouse

632T5(TAN), 633T5(AQUA), C.1957
PLASTIC, 5 TUBES, 1 BAND
TAN $40
AQUA $50

Westinghouse

640T5(BROWN), 641T5(AQUA), C.1957
PLASTIC, 5 TUBES, 1 BAND
BROWN $40
AQUA $50

Westinghouse

644T5(BROWN), 645T5(AQUA), C.1957
PLASTIC, 5 TUBES, 1 BAND
IVORY $40
RED $65

Westinghouse

451T5(BROWN), 452T5(IVORY),
453T5(GREEN), 454T5(BLUE), C.1954
PLASTIC, 5 TUBES, 1 BAND
BROWN, IVORY $45
GREEN, BLUE $75

Westinghouse

461P4(RED), 462P4(IVORY),
463P4(TAN), 464P4(GREEN), C.1954
PLASTIC 4 TUBES, 1 BAND
IVORY, TAN $75
RED, GREEN $110

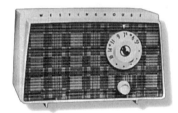

Westinghouse

H499TB(BLACK), H500TB(RED),
H501TB(BROWN), C.1957
PLASTIC, 5 TUBES, 1 BAND
BLACK, BROWN $45
RED $80

Westinghouse

H504P4(MAROON), H505P4(BLACK),
H506P4(GREEN), H507P4(RED),
C.1957
PLASTIC, 4 TUBES, 1 BAND
MAROON, BLACK $75
RED, GREEN $125

Westinghouse

H519P4(MAROON), H520P4(RED),
H521P4(TAN), C.1955
PLASTIC, 4 TUBES, 1 BAND
MAROON, TAN $60
RED $90

Westinghouse

523T4(BROWN), 524T4(IVORY),
525R4(RED), C.1955
PLASTIC, 5 TUBES, 1 BAND
BROWN, IVORY $45
RED $75

Westinghouse

H526T5A, C.1956
PLASTIC, 5 TUBES, 1 BAND
$40

Westinghouse

H536T6, C.1956
PLASTIC, 6 TUBES, 1 BAND
$45

Westinghouse

H538T4(BLACK), H539T4(IVORY),
H540T4(RED), C.1955
PLASTIC, 4 TUBES, 1 BAND
BLACK, IVORY $75
RED $145

Westinghouse

H562P4(TAN/BROWN),
H563P4(GREY/BLACK),
H564P4(WHITE/TAN), C.1957
PLASTIC, 4 TUBES, 1 BAND
$75

Westinghouse

H686P6, c.1960
PLASTIC, 4 TUBES, 1 BAND, DC
PINK $85

Westinghouse

WR20, c.1935
WOOD, 4 TUBES, 1 BAND
$90

Westinghouse

WR21, c.1935
WOOD, 5 TUBES, 1 BAND
$90

Westinghouse

WR22, c.1934
WOOD, 5 TUBES, 1 BAND
$110

Westinghouse

WR23, c.1935
WOOD, 7 TUBES, 2 BANDS
$225

Westinghouse

WR27, c.1935
WOOD, 5 TUBES, 1 BAND
$90

Westinghouse

WR120, c.1938
BAKELITE, 5 TUBES, 1 BAND
$65

Westinghouse

WR166, c.1939
BAKELITE, 5 TUBES, 1 BAND
$225

Westinghouse

WR168, c.1939
WOOD, 5 TUBES, 1 BAND
$50

Westinghouse

WR169, c.1939
WOOD, 5 TUBES, 2 BANDS
$60

Westinghouse

WR170, WR270, c.1939
WOOD, 5 TUBES, 2 BANDS
$65

Westinghouse

WR172, WR272, c.1939
WOOD & METAL
6 TUBES, 2 BANDS
$90

Westinghouse

WR222, C.1938
WOOD, 5 TUBES, 2 BANDS
$65

Westinghouse

WR224, C.1938
WOOD, 5 TUBES, 2 BANDS
$65

Westinghouse

WR256, C.1939
WOOD, 5 TUBES, 1 BANDS
$50

Westinghouse

WR258, C.1938
WOOD, 6 TUBES, 2 BANDS
$75

Westinghouse

WR274, C.1939
WOOD, 7 TUBES, 3 BANDS
$95

Westinghouse

WR288, C.1940
WOOD, 8 TUBES, 3 BANDS
$135

Westinghouse

WR470, RADIO-PHONO, C.1940
WOOD, 5 TUBES, 1 BAND
$60

Westinghouse

WR473, RADIO-PHONO, C.1940
WOOD, 7 TUBES, 2 BANDS
$55

Westinghouse

WR478, RADIO-PHONO, C.1941
WOOD, 5 TUBES, 1 BAND
$50

Westinghouse
CANADA

5T121, C.1957
PLASTIC, 5 TUBES, 1 BAND
BLACK, BROWN, IVORY, GREY $70

Westinghouse
CANADA

5T124, C.1957
PLASTIC, 4 TUBES, 1 BAND
BLACK, BROWN, IVORY, GREY $85
GREEN $150

Westinghouse
CANADA

5T125, C.1957
PLASTIC, 5 TUBES, 1 BAND
BLACK, BROWN, IVORY $45
RED, GREEN $80

Westinghouse
C A N A D A
5T126, C.1957
PLASTIC, 5 TUBES, 1 BAND
TAN, BLACK $40
BLUE $50

Westinghouse
C A N A D A
100X, C.1940
WOOD, 6 TUBES, 2 BANDS
$80

Westinghouse
C A N A D A
501M, C.1950
PLASTIC, 4 TUBES, 1 BAND
IVORY $90
BLUE $200

Westinghouse
C A N A D A
507, C.1958
BAKELITE, 5 TUBES, 1 BAND
$30

Westinghouse
C A N A D A
604, C.1949
BAKELITE, 6 TUBES, 3 BANDS
$55

Westinghouse
C A N A D A
6T116, C.1956
WOOD, 6 TUBES, 1 BAND
$125

Westinghouse
C A N A D A
X2395-11, C.1958
PLASTIC, 5 TUBES, 1 BAND
$45

Wilcox-Gay
3T66, 'COMPANIION, C.1934
WOOD, 6 TUBES, 1 BAND
$140

Wilcox-Gay
4CD5-29, C.1935
WOOD, 5 TUBES, 1 BAND
$750

Wilcox-Gay
A-32, C.1938
WOOD, 7 TUBES, 2 BANDS
$250

Wilcox-Gay
A-54, C.1939
WOOD, 7 TUBES, 3 BANDS
$140

Wilcox-Gay
A-43, 'WAL-RADIO', C.1939
WOOD, 5 TUBES, 1 BAND
$110

Wilcox-Gay

A-50, C.1939
WOOD, 5 TUBES, 1 BAND
$65

Wilcox-Gay

A-54, C.1939
WOOD, 7 TUBES, 2 BANDS
$140

Wilcox-Gay

A-56, PHONO, C.1939
METAL, PHONO REMOTE
$30

Wilcox-Gay

A-57, PHONO, C.1939
WOOD & METAL, PHONO REMOTE
$40

Wilcox-Gay

A-76, C.1941
WOOD, 5 TUBES, 2 BANDS
$40

Wilcox-Gay

'CARDINAL', C.1934
WOOD, 5 TUBES, 1 BAND
$750

Wilcox-Gay

'CHAMELEON', C.1938
METAL, 5 TUBES, 1 BAND
$225

Wilcox-Gay

'VOGUE', C.1935
WOOD, 5 TUBES, 1 BAND
$225

WINGS

588, C.1937
WOOD, 5 TUBES, 1 BAND
$140

WRL

KIT, C.1944
BAKLEITE, 3 TUBES, 1 BAND
$65

ZANEY-GILL

'CLARION', C.1930
WOOD & POT METAL, 5 TUBES, 1 BAND
$600

ZANEY-GILL

MIDGET, C.1932
WOOD, 4 TUBES, 1 BAND
$225

ZANEY-GILL

'Music Box', c.1931
WOOD, 5 TUBES, 1 BAND
$325

ZANEY-GILL

'Vita-Tone', c.1931
WOOD, 5 TUBES, 1 BAND
$775

Zenith

6G001Y, c.1946
CANVAS
$75

Zenith

8G005Y, 'Transoceanic', c.1946
CANVAS, 8 TUBES, 4 BANDS
$175

Zenith

580, RADIO-PHONO, c.1941
WOOD, 6 TUBES, 2 BANDS
$110

Zenith

602, RADIO-PHONO, c.1939
WOOD, 6 TUBES, 1 BAND
$175

Zenith

B615F, c.1959
PLASTIC, 5 TUBES, 1 BAND
$20

Zenith

F508A, c.1958
PLASTIC, 5 TUBES, 1 BAND
$20

ZEPHYR

#1, c.1960
PLASTIC, 4 TUBES, 1 BAND
WHITE $40
PINK $60
BLUE $70

Zephyr

41X6, c.1938
WOOD, 5 TUBES, 1 BAND
$175

Zephyr

23A5, c.1938
WOOD, 5 TUBES, 2 BANDS, DC
$50

Zephyr

25B5, c.1938
WOOD, 5 TUBES, 2 BANDS, DC
$50

ZEPHYR
25X7, C.1938
WOOD, 7 TUBES, 2 BANDS
$110

ZEPHYR
30X5, C.1938
WOOD, 5 TUBES, 1 BAND
$70

ZEPHYR
31X5, C.1938
5 TUBES, 1 BAND
$30

ZEPHYR
31YP6, RADIO-PHONO, C.1938
WOOD, 6 TUBES, 2 BANDS
$100

ZEPHYR
32P6, C.1938
6 TUBES, 3 BANDS, DC
$60

ZEPHYR
33P6, C.1938
6 TUBES, 3 BANDS, DC
$45

ZEPHYR
35X5(BLACK) 36X5(IVORY), C.1938
5 TUBES, 2 BANDS
BLACK $200, IVORY $275

ZEPHYR
42X6, C.1938
WOOD, 6 TUBES, 1 BAND
$175

ZEPHYR
43X5, C.1938
WOOD, 5 TUBES, 2 BANDS
$150

ZEPHYR
61X8, C.1937
WOOD, 8 TUBES, 3 BANDS
$235

ZEPHYR
62X8, C.1938
WOOD, 8 TUBES, 3 BANDS
$225

ZEPHYR
67Y8, C.1938
WOOD, 8 TUBES, 3 BANDS
$265